Scientism and Technocracy in the Twentieth Century

Scientism and Technocracy in the Twentieth Century

The Legacy of Scientific Management

Richard G. Olson

LEXINGTON BOOKS
Lanham • Boulder • New York • London

Published by Lexington Books
An imprint of The Rowman & Littlefield Publishing Group, Inc.
4501 Forbes Boulevard, Suite 200, Lanham, Maryland 20706
www.rowman.com

Unit A, Whitacre Mews, 26-34 Stannery Street, London SE11 4AB

British Library Cataloguing in Publication Information Available

Library of Congress Cataloging-in-Publication Data

Names: Olson, Richard, 1940- author.
Title: Scientism and technocracy in the twentieth century : the legacy of scientific management / Richard G. Olson.
Description: Lanham : Lexington Books, [2015] | Includes bibliographical references and index.
Identifiers: LCCN 2015037346| ISBN 9781498525701 (cloth : alk. paper) | ISBN | ISBN 9781498525718 (electronic)
Subjects: LCSH: Technocracy--History--20th century. | Scientism--History--20th century. | Management science--History--20th century. | Technology and state--History--20th century.
Classification: LCC HB87 .O47 2015 | DDC 338.9/26--dc23 LC record available at http://lccn.loc.gov/2015037346

∞ ™ The paper used in this publication meets the minimum requirements of American National Standard for Information Sciences Permanence of Paper for Printed Library Materials, ANSI/NISO Z39.48-1992.

Printed in the United States of America

Contents

counted as technocrats, but they are probably the most influential among those at high government levels whose power derives from technical knowledge. In addition, technically trained economists have played a huge role in EU policy, and since the 1960s economists have replaced engineers and physicists as the dominant group within such paradigmatic technocratic organizations as think tanks including the RAND corporation. (See chapter 6, pp. 148–150).

There is some indication that economists who have served governments in developing countries in recent years have not shared the sympathy for democratic, or "debating society" politics shown in the 1970s by economists in advanced industrial societies, nor have they demonstrated the same concern with the distribution of goods. (See chapter 5, pp. 128–129). In fact, I will argue in chapter 6 that recent developments in engineering and economics education indicate that as engineers seem to be becoming less technocratic by Putnam's standards, economists seem to be coming more so.

I will usually use the term "technocrat" in the very simple way identified by Putnam and Jacques Menaud—that is, to indicate *one who exercises power by virtue of their technical knowledge*—and I will usually use the term "technocracy" to indicate any social unit in which technocrats play a large role in determining major policies, in deciding how those policies should be implemented, or both. Each term has been used in the literature in a number of different but related ways, and in certain contexts I will use them differently; though always making those divergences clear. Consider just two examples by way of illustration. First, there is a convention among students of Chinese politics to limit the term "technocrat" to a person who has a higher degree in a technical field—typically engineering or agronomy—and has professional experience in that field before accepting an appointment as a public official. In many, but not all, cases their political appointments are directly related to the technocrat's previous professional experience.

The major reason that I am interested in technocrats is that they constitute what recent scholars have called an "epistemic community," that is they share a wide range of attitudes and assumptions about such ideas as " knowledge," "science," "objectivity" "subjectivity," "rationality," "efficiency," "politics," and "democracy"—attitudes and assumptions that have an effect upon their public actions. So even if a technically trained Chinese official holds a position and thus has authority because of family connections rather than because of special expertise, I will use the term technocrat because of the assumption (only occasionally directly verifiable) that their training and experience shapes the way they make decisions and/or the content of those decisions.

Second, many persons, including myself, are interested in exploring the relationships between technocratic decision making and political, and/or democratic decision making. Many possibilities exist. One of the most common follows an argument made by Max Weber, Woodrow Wilson, Frank Goodnow, and others who saw the growing complexities of the modern state as leading to the growth of bureaucracies in which expertise became the primary criterion for appointment and in which decisions were usually made in a way that I have been calling technocratic. Weber, Wilson, and Goodnow were concerned about the growing role of technocratic decision making and all three insisted upon a separation between the *making* of public policies, which would focus on goals and would be carried out politically, and the *implementation* of those policies which would be carried out by bureaucrats/ technocrats.

A few contemporary political theorists and political scientists—especially those who are associated with neo-liberalism—continue to hold this position. They persist in believing that there is a fairly clear line between political decisions, which should respond to public opinion through some mechanism, whether it is direct democracy, representative democracy, or simply toleration of the current dominant party or dictator, and administrative decisions about the implementation of such policies. Curiously, even some of the most respected contemporary Science and Technology Studies students, including Harry Collins, seem to be returning to this view; thus he and Robert Evans write, "Political choice is a domain which we do not consider to be technical. Failing to maintain a distinction between science and technology, on the one hand, and politics on the other leads to the stark choice between technological populism, in which there are no experts, and fascism, in which the only political rights are those gained through supposed technical expertise" (2007, 8).

From my perspective, policies and implementation strategies are often so intimately intertwined that the classical assumption of separation is virtually impossible to sustain. It is also true that political choices are often made between options that have been formulated by technical experts, so the terms and limits of political discussion are restricted to fit within perspectives dominated by technocratic views and vocabularies—in Herbert Simon's terms, the choices are restricted to lie within a technocratically bounded rationality, or in terms recently used by a group of historians including Paul Erickson, Judy L. Klein, Lorraine Daston, Rebecca Lemov, Thomas Sturm, and Michael Gordin to characterize late-twentieth-century developments, they are restricted to accepting "Cold War rationality" (Erickson et al. 2013, 3–4)." There may, however, be contexts in which I have to acknowledge a perceived radical separation between political policies and technocratic implementations because all of the historical actors being discussed did so.

Finally, much recent history and philosophy of science acknowledges that the long-standing claims of rationality and objectivity on the part of scientists and engineers are no longer defensible in traditional ways and that evidentiary standards, for example, are more matters of social convention than of simple logic. So technocratic decision making is vastly more complex than it was widely thought to be before the 1960s (Collins and Evans, 2007). While I accept this revisionist position, few if any technocrats, even today, allow it to undermine their confidence in the objectivity and rationality of science or the fundamental superiority of scientifically grounded decisions to politically negotiated ones. RAND corporation, for example, continues to argue that its policy recommendations are "disinterested and objective" (RAND 2009). In what follows, I will remind readers of the mismatch between technocratic beliefs and social reality when the story demands that I do so.

Science and Scientism

Back in 1982, when I began the project which has become a study of scientism, I offered provisional definitions of science and scientism which I will continue to use in this work. "Science," I wrote,

> is taken to be a set of activities and habits of mind aimed at contributing to an organized , universally valid, and testable body of knowledge about phenomena. At any given time and place, these general characteristics are usually embodied in systems of concepts, rules of procedure, theories, and/or model investigations that are generally accepted by groups of practitioners—the scientific specialists (Olson 1982, 7–8).

At least three comments should be made about this definition. First, I am virtually certain that no explicit definition of the term "science" can satisfy everyone. On the one hand, at any time and place there are many sciences rather than one, each of which may have somewhat different concepts and procedural rules than the others, so that specialists from different fields may disagree about the boundaries of "science." Perhaps more importantly, virtually every definition that I have encountered has been designed to serve the interests of some particular group. Second, the fact that I claim that scientists seek to discover universally valid and testable knowledge does not mean that such knowledge can ever be found. Unattainable goals like the search for the Holy Grail or the search for immortality may very well direct the behaviors of people as powerfully as attainable ones. Finally, the admission that groups of scientific practitioners collectively determine what concepts and procedures are acceptable opens a wide door for societal influences to play a major role in the sciences. As I argued in *Science and Scientism in Nineteenth-Century Europe,* there was a time during the early nineteenth century in Germany when *Naturphilosophie*, a set of concepts and practices that were grounded in German idealist philosophy and which violated most nineteenth- and twentieth-century understandings of science elsewhere, were widely accepted among both physicists and physiologists; yet I argued that many of those who were self-styled *naturphilosophen*, were genuine scientists (Olson 2008, 87–111).

"Scientism" I take to be any attempt to apply the concepts, practices, and attitudes associated with one or more of the natural sciences at any time and place to new domains, especially domains related to human interests and activities. The implicit assumption that human interests and activities are in some sense not "natural," which can be read from the previous sentence and which is held by some, but by no means all, social scientists today, embeds a distinction made already by Aristotle when he distinguished between theoretical sciences, the claims of which were *always* true because the objects of concern were not subject to human influence or choice, and practical sciences, like politics and law, whose claims were only *usually* true because their objects could be modified by human interests and choices.

My personal view is that the Aristotelian distinction is very important and legitimate, because I persist in the view that I and other humans are capable of real choices which are not simply illusionary. Thus I tend to think that the social or human sciences are not as deterministic as the non-human natural sciences. This view, however, does not deny that there can be legitimate social or human sciences. So in *Science and Scientism in Nineteenth-Century Europe*, I argued that Saint Simon's attempt to construct a social science was both scientistic and scientific because it sought to extend ideas and practices from the natural sciences to social issues and because it did so in a way that did not violate the integrity of the scientific ideas and practices common in the early nineteenth century. Indeed, I argued that Saint Simon's reflections on current scientific ideas and practices were more sophisticated than those of the natural scientists from whom he borrowed (Olson 2008, 59–61). The fact that he argued scientifically, of course, does not mean that his claims were correct by later standards just as the fact that advocates of Ptolemaic astronomy or phlogiston chemistry argued scientifically makes their claims correct by later standards. It is critical to keep in mind that a genuinely scientific claim is not necessarily a true claim, especially to the extent that one holds some correspondence theory of truth.

It will be the overall argument of this book that scientistic and technocratic attitudes and arguments have dominated not only factory level and business level decisions but also public policy decisions throughout the world during the vast bulk of the twentieth century. Many of those attitudes developed their public authority as a consequence of their embodiment within the Scientific Management movement promoted by Frederick Winslow Taylor and his close colleagues. These attitudes and arguments have had very different consequences depending on the specific local cultural circumstances into which they were introduced. Under most conditions they have played a central role in increasing the production of material goods and the effective providing of public services. At the same time, however, because of their relatively narrow focus and their lack of interest in distributive justice, scientism and technocracy have tended to exacerbate income inequalities in the world and to favor easily quantifiable considerations over those not so easily quantified. Furthermore, the extreme sensitivity of technocratically driven decisions to the quality of information inputs, coupled with a relative naivete among technocrats regarding the honesty of data collectors who are at the same time political actors, has led to frequent cases of bad policy advice.

I am convinced that we live in a world in which public decisions unavoidably depend to a large extent on technically competent advisory input; but I am equally convinced that unless the foundations of technical advice are broadened well beyond what they incorporate today, that input is as likely to diminish the quality of our public and private lives as to enhance their quality.

Acknowledgments

I owe special thanks to four Claremont Graduate School students: Xaiolong Wang, whose paper on changes in Economics education in the United States during the last century was the source of most of what I say about that topic in chapter 6; Sean Butler, who did the analysis of the RAND staff based on the RAND website in 2013; and Ruben Lopez and Bonnie Hardman, who, along with Sean and Xaiolong, read drafts of all chapters and gave me feedback on them. My history colleague at Harvey Mudd College, Hal Barron, read a version of the manuscript and made a number of suggestions, some of which I accepted as did my STS colleague, Marianne de Laet at HMC, Dan Segal at Pitzer College, and Juan Lucena, who I got to know while visiting for the 2002–2003 academic year at Colorado School of Mines. My engineering colleague Ruye Wang, who teaches during the summers in Bejing, read through portions of chapter 5 on China in an attempt to keep me honest.

Though Judith Merkle has been gone for several years, I owe her for stimulating my initial interest in Taylorism and its ideological impact outside of the United States. We were among the founders of the Claremont College's STS program, we audited one another's classes, and I miss her quirky character. Though we come from very different perspectives, conversations with Robert Klittgard, former dean of the Pardee RAND Graduate School, former president of Claremont Graduate University, and longtime advocate of transdisciplinary policy studies, have enriched my understanding substantially, though he certainly would not agree with much that I have to say about transdisciplinarity.

I benefited tremendously from my interactions with the participants in a conference on Engineers and the Political Process sponsored in 2012 by the Hixon Forum for Responsive Technology at Harvey Mudd College. These included Derin Barney; Mark Brown; Jason Delborne; Marianne de Laete;

Captain Mark Haggerott, USN; Juan Lucena; Amy Slaton; Pete Suttmeyer; Zuoyue Wang, and Matthew Wisnioski.

The editorial staff at Lexington Books, Brian Hill, Brighid Stone, and Eric Kuntzman, have been helpful and encouraging, and I owe their anonymous referee for making a number of valuable suggestions. The image of the statue from the fountain at the entrance to the Science Building at the 1933 Chicago Century of Progress Exposition as well as permission for its use was provided by the Library of the University of Illinois at Chicago, and the picture of the Esplanade of the Ministries in Brasilia is used by permission of the photographer, Vanessa Volk. As usual, my wife, Kathy Collins Olson, provided unending support and our Corgi, Parker, kept me entertained and exercised.

Richard Olson
Claremont, California
August 2015

Chapter One

Scientism among the Engineers

Origins of the Scientific Management Movement

In *The Practice of Management* (1954), Peter Drucker carried the most progressive and expansive visions of Fredrick Winslow Taylor, the famous and notorious father of Scientific Management, into the late twentieth century, maintaining that Taylor, rather than Karl Marx should be included in the triumvirate along with Charles Darwin and Sigmund Freud as "makers of the modern world" (Kanigel 1997, 501). He wrote, "Scientific Management . . . may well be the most powerful as well as the most lasting contribution that America has made to Western thought since the Federalist Papers" (Drucker 1954, 281). Moreover, claimed Drucker, "in modern society there is no other leadership group but managers. If the managers of our major institutions, and especially of business, do not take responsibility for the common good, no one else can or will" (Drucker 1973, 325).

During the 1960s John Kenneth Galbraith documented the extent to which expert managers had replaced the ownership of corporations as the primary decision makers in our businesses and our society in *The New Industrial State* (1967). And the early twenty-first-century economic disruption of the American economy as a consequence of the failure of the managers of American financial institutions to take responsibility for the common good in the face of pressures to generate short-term gains for their investors has seemed to reinforce Drucker's claim that no one else appears to have the ability or the will to protect the public. I do, however, have at least one major criticism of Drucker's claims, and that is that he restricted them to "Western thought"; for Scientific Management and the rise of an expert managerial class has had a major impact, both directly and indirectly, on Eastern thought—especially in India, Japan, Korea, and China—and in many nations of the south as well.

For present purposes, Scientific Management is important as one of the most significant vehicles for carrying an essential dimension of scientism—a supreme confidence that the practices and attitudes associated with the natural sciences could be effectively used to solve a virtually unlimited range of social problems as well—throughout the globe during the early twentieth century.

The Origins of Scientific Management

What came to be known as Scientific Management grew out of a movement in the last two decades of the nineteenth century in the United States, largely among mechanical engineers. The movement, associated initially with the term "systematic management" (Litterer, 1961, 461–476), involved an ensemble of relatively modest attempts to improve the efficiency of shop practices in manufacturing industries. Such attempts involved the design of new, specialized machines, the use of time and motion studies to decrease wasted effort, the reorganization of work flow, incentive systems to increase employees' time on task, new cost accounting procedures, and attempts to match workers capabilities to specific tasks. According to its practitioners, the great virtue of the approach which they took to increasing efficiency was that in it "scientific laws and rules are substituted for the old-fashioned individual initiative of the men" (Layton 1986, 141).

In 1903, Frederick W. Taylor synthesized the work of many of his colleagues and added notions that he had developed in connection with his attempts to improve efficiency at Midvale Steel, Bethlehem Steel, and the Simonds Rolling Machine Company in *Shop Management*, a work which had modest success among American engineering managers before 1911, when it became an international sensation for reasons that we will soon discuss. Taylor insisted that his primary goal was to promote both high wages and low labor costs, thus benefiting both workers and owners (Taylor 1919, 22).

Taylor was appalled at what he saw as the conflict between workers and owners in factory settings, with a consequent negative consequence for productivity. Owners tended to try to maximize their profits by minimizing labor costs through keeping pay rates as low as possible. In response, workers sought to maximize their income by "soldiering"—that is, by lowering their productivity through work slowdowns. Any worker who actually made an effort to boost productivity soon incurred the wrath of coworkers and was punished appropriately. As long as owners and workers faced one another primarily as antagonists, productivity was kept low and neither group benefited. Taylor's stated goal was to increase productivity and to "secure the maximum prosperity for the employer, coupled with the maximum prosperity for each employee" (Taylor 1911, 9).

Though he admitted that several different kinds of incentive plans, including profit sharing and wages-plus-bonus systems, could be of some benefit in increasing the productivity of workers, Taylor outlined the weaknesses of previous incentive structures and promoted what he called a "differential piece-rate system" that set productivity standards "scientifically" by dividing tasks into their smallest elements, timing each element as done by the best workers using a stopwatch, and then summing up the times and adding a percentage for rests and for unavoidable delays. Details of how to measure the unit times and formulae for calculating the maximum daily outputs per worker developed by Taylor's colleague, Sanford E. Thomson, were provided in *Shop Management* (Taylor 1919, 149–176). The standard daily goal for productivity was then set slightly below the maximum, and any worker who met the goal received a daily wage that was the norm in the industry plus some percentage that depended upon the kind of task being done. Taylor claimed that the percentage of overpay required to motivate workers of different kinds to work at their daily maximum was also determined empirically, though he provided no evidence regarding this claim (Taylor 1919, 26).

Taylor's system clearly offered substantially greater pay to those "first class men" who met his standards; but he recommended making the tasks "so severe that not more than one of five laborers could keep up" (Taylor 1919, 54–55), and he insisted that in order to keep them motivated to achieve their high level of productivity it was desirable, "for the workmen quite often to have the object lesson of seeing men laid off for failing to earn high wages and others substituted in their places" (Taylor 1919, 74). Such views unsurprisingly tended to alienate many workers and virtually all union leaders. Taylor responded by arguing that there was a huge range of work tasks and that virtually anyone could become a first class man at some kind of work. Furthermore, he railed against union insistence on equal hourly pay for all workers doing the same kind of work on the grounds that "men are not born equal, and any attempt to make them so is contrary to nature's laws" (Taylor 1919, 190). On the other hand, he admitted that when employers trampled employee rights and when *employers* offered no incentives for better workers, then unionization might be necessary (Schachter 1989, 55–56).

Although Taylor proposed setting high productivity standards based on the short-term speed of expert workers, he did admit that workers new to any task should be given a significant period to achieve those standards, he provided for experts to train new recruits, and he insisted both that time be allowed for machine maintenance and repair and that materials and proper tools be provided the worker in a timely fashion so the worker was not held accountable for the failures of others. When others tried to use Taylorite methods of setting productivity standards without providing the necessary support systems, they almost universally failed to reach their goals.

To increase productivity, Taylor and his contemporaries also designed new specialized tools and studied the optimum speeds, depth of cuts, and cutting tool shapes and tool materials for metal cutting and shaping. Since maximum productivity in metal working depended on precision setting of machines as well as precise movements in accomplishing tasks, Taylor argued that workmen should have little, if any, flexibility in choosing how to undertake their work. A planning department would provide a written card giving the exact details of every step in accomplishing any task. "This complete standardization of all details and methods is not only desirable but absolutely indispensable as a preliminary to specifying the time in which each operation should be done, and then insisting that it should be done in the time allowed," wrote Taylor (1919, 123). Finally, Taylor argued that the planning department should be involved in all phases of production, setting the times for each task as well as the compensation levels, arranging for the timely purchase of materials and machines, the scheduling of maintenance, the hiring of workmen appropriate for each job, the flow of work, and the evaluation and disciplining of workmen, all of which required detailed record keeping.

Within the planning department, whose members Taylor designated as "non-productive" employees, he argued that the principle of "functional management" should replace the old hierarchical tradition of military organization (Taylor 1919, 99). Thus, each worker, rather than connecting with a single representative of management, would be directed, in Taylor's illustrative case, by "eight different bosses, each of whom performs his own particular function" (1919, 99). Some of these bosses—those in the planning department—would ordinarily communicate through written instructions placed on cards given to the workman to explain how each task was to be done. Other bosses—those on the shop floor—would be in direct contact with the productive employees, helping the workmen to meet their productivity goals. One would see to it that materials and special tools were ready at all times, a second would see that the proper tools and speeds and feeds were being used, a third would inspect the products for quality, and a fourth would see that the worker maintained his machine properly. Each boss would keep a daily record of each worker's performance in his area and forward it to the appropriate planning department member who would provide daily feedback to the worker regarding his productivity. When any worker was having difficulty, the appropriate floor boss would provide instruction to bring him up to the standard required (Taylor 1919, 98–111). Short-term failure to meet standards would be punishable by fines. Long-term failures would lead to dismissal.

Though more than half of *Shop Management* focused on details of the differential piece-rate system for motivating workers to operate at maximum efficiency, Taylor offered a brief—and crude—discussion of how to choose the correct kind of workers for different kinds of jobs (Taylor 1919, 140–143). He also emphasized the standardization of tools as well as procedures. He discussed a set of experiments on hardening tool steel which he had done with J. Maunsel White that had led to the creation of Taylor-White steel, which was capable of increasing cutting speeds by 300 percent (Taylor 1919, 124–125). Finally, he described a set of experiments carried out over a nine-year period on the proper care and tightening of the belting used to drive machinery, which doubled the average pulling power of belts by reducing slippage and halved the downtime of equipment due to belt failure (Taylor 1919, 125–126).

In *Shop Management*, based on his personal experience, Taylor warned owners that transition from traditional management to Scientific Management would take time and substantial amounts of money; it would certainly be opposed by workers initially; and counterintuitively, as the implementation of Scientific Management was more complete it would lead to an increasing ratio of nonproductive to productive workers—that is, the planning department would become a greater part of the entire enterprise. In addition, he argued that Scientific Management should be incrementally established, starting with one or a very few tasks. But, he argued, using the Midvale Steel, Bethlehem Steel, and Simonds Rolling Machine Company examples, long-term reductions in labor costs coupled with increased productivity would make the transition well worth the difficulties that had to be faced.

While some owners had the patience to weather transitional difficulties, many were driven by the demands for short-term profits to give up; so Scientific Management was not rapidly and widely adopted in American industrial settings until the demand for military goods during WWI forced industries to emphasize efficiency. When that happened, the differential piece-rate system was usually the last feature of Scientific Management implemented if it was implemented at all.

The legitimacy of the claims by Taylor and other advocates of Scientific Management to being "scientific" has been challenged frequently, in part because after carefully measuring the time taken by expert workers to accomplish some task, management experts typically used arbitrary or "conventional" multiplication factors to set productivity standards (Kanigel 1997, 13). But in doing so, Scientific Management advocates were acting like responsible engineers, who after scientifically calculating the load born by some structural element, designed the element to be capable of sustaining some conventional multiple of that load to ensure safety. Such practices, though grounded in scientific measurements or calculations, also usually incorporated disciplinary conventions. In addition, there is no doubt that Taylor did either conduct or have conducted careful sets of experiments on air-hardened cutting steel, belting, and the optimal cutting speeds for various machines, as well as studies on the varying capacity of ball-bearing inspectors to detect flaws under a variety of conditions. In the latter case he made sure that the evaluators of different inspectors were doing "blind" evaluations to avoid personal bias (Taylor 1919, 87–91).

While it would certainly be wrong to claim that Taylor and his close colleagues always adhered to the best practices for doing scientific studies and reporting their results, and while some of Taylor's claims to being scientific were no doubt exaggerated, on balance his promotion of scientific methods to guide industrial practices was legitimately grounded in his own work. The one major exception was his discussion of matching workers to jobs, which was being done in much more sophisticated and "scientific" ways by psychologists in personnel departments while Taylor promoted traditional, rule of thumb, methods.

Regardless of his own occasional lapses and exaggerations, Taylor clearly and effectively communicated to his colleagues and readers an ideology of careful and patient experimentation. Writing to Morris Cooke, his follower and sometime collaborator, he stated his view:

> The true experimenter must be an enthusiast; he should have the keen delight in obtaining a result that the ordinary man would have in finding a diamond mine. This very enthusiasm, however, leads most experimenters into perhaps their worst error or fault, namely a desire to always get a positive, useful result, whereas perhaps nine out of ten experiments when truthfully carried out must lead to negative results. . . . The true experimenter sets no time limit, but is willing to look forward for years and plod away. (Schachter 1989, 27)

At least in its early versions, Scientific Management ignored social and other potentially significant complicating environmental factors such as temperature and humidity when determining productivity standards; but once again the *ceteris paribus* assumption was common in all sciences, and when environmentally significant factors became known, they were often incorporated into experiments and calculations (Kanigel 1997, 332).

Virtually all of the early practitioners of Scientific Management with the exception of ex-brick layer, Frank Gilbreth, were from upper-middle-class backgrounds; and with few exceptions they agreed with Taylor in viewing ordinary laborers as undermining efficient production through "soldiering" or intentional work slow-downs. By pushing various techniques to speed up the pace of work, virtually all of these proponents of efficiency earned the enmity and opposition of the U.S. labor movement, in large part because employers had traditionally done what they still do to salesmen. Once workers met some new quota for getting some kind of bonus, employers simply increased the quota. It is hardly surprising, then, that workers were suspicious of Taylor's claim that "scientific rate setting" precluded this traditional practice, especially since work speeds could and would be reestablished whenever new machines or procedures were introduced. Moreover, in spite of claiming concern for workers, most early Scientific Management experts showed little sympathy for the plight of unskilled laborers, who naturally resented losing control over their work and the loss of their jobs if they could not meet the standards set by management.

We shall soon see that in his most famous work Taylor reflected this middle-class and college-educated view; but in his consulting practices, which were often reflected more accurately in his technical papers and speeches, he showed much greater respect for working mechanics and their knowledge. As an apprentice pattern maker, he wrote, "I there learned appreciation, respect, and admiration for the everyday mechanic," and later in his career, speaking to The Society for the Promotion of Engineering Education, he urged college students to spend a year working in a factory after their freshman year, where, "they would learn that lathe operators were their equals," and would discover that, "the work machinists did required a fund of knowledge equally as important as that required in the traditional professions" (Schachter 1989, 21).

In spite of his belief that his time studies based on the analysis of motions done by expert workmen would lead to maximum efficiency and his insistence that workers initially follow exactly the directions of management experts, Taylor did encourage workmen to subsequently suggest changes in procedures or machines that might improve efficiency, and he admitted that listening to workmen was an important factor in maintaining good morale in a shop, which, in turn, increased productivity (Taylor 1919, 80). In a speech to the City Club of Philadelphia he described his procedures for soliciting worker suggestions at Link Belt Manufacturing and for rewarding those whose suggestions were implemented by attaching their names to the new procedures. Cooke reiterated this position in his 1913 essay on "The Spirit and Social Significance of Scientific Management," writing that, "We feel that everybody in the establishment, from the office boy up—and those out of the establishment for that matter—should be given the widest opportunity for making suggestions and criticisms. [And] suggestions that are in any way adopted should be fairly compensated for" (492–493). Taylor also assumed that most managers would be promoted up from shop floor positions as they demonstrated abilities to do what he considered to be higher level work.

Finally, though many of those who sought to implement Taylorite methods failed to do so, Taylor typically held open meetings with workers before starting to implement his plans to explain how and why Scientific Management would benefit them; to allow them to ask questions; and to respond to potential objections (Schachter 1989, 52). Those who paid attention to his personal practices and minor writings realized that Taylor was less authoritarian and more sympathetic to labor than his more popular presentations suggested.

Taylor never renounced his general opposition to unions. But after Taylor's death in 1915 a number of his relatively orthodox followers who had established the Taylor Society joined with Progressive advocates of "industrial democracy"—a movement in which factory policies were established jointly by management experts and workers—without feeling that they were doing serious violence to his principles. Indeed, by 1919 the Taylor Society had become a primary center for the promotion of industrial democracy (Haber 1964, 129). Shortly thereafter a few union leaders, especially in the garment trades, even began to argue that their unions should embrace Scientific Management as a means to increase wages and fight capricious and unfair labor practices on the part of owners (Haber 1964, 150).

Taylor and his followers often managed to irritate and alienate their business clients as well as union leaders—not just because the introduction of Scientific Management practices initially cost time and money, but also because they promoted much higher pay rates than industry averages. Owners were rarely willing to accept the notion that such high wages were critical motivators, nor were they enthusiastic about turning control over their businesses to a bunch of outsiders.

Taylor had made his differential piece rate system grounded in scientific time-studies of work processes the centerpieces of *Shop Management*, and his foreign followers, as we shall see in chapter 3, often sought to implement this aspect of Scientific Management first, in part because it demanded very little capital outlay. But his American disciples found that they had the greatest successes in implementing Taylorite management schemes, as Taylor and his coworkers had done at the model factories of Link-Belt and Tabor Manufacturing, when they began by focusing on machinery—adjusting the speeds and cut depths of lathes, for example—, improving the condition of major machinery, and providing plentiful supplies of smaller tools (Nelson 1980, 147). Next, they introduced new tool maintenance procedures, modified methods of handling materials, and initiated some forms of functional management. Then they created a planning department and reorganized the stockroom, tool room, and accounting procedures "so that the men would find their work not only profitable, but pleasant" (Nelson 1980, 145). Only after gaining the trust and cooperation of workers did they establish the differential piece rate system based on time studies at Link-Belt and Tabor Manufacturing.

In almost a third of the cases connected to American private firms, the initiation of Scientific Management, though successful in increasing production while controlling labor costs, was terminated before the differential piece rate system was implemented because opposition was so strong, and Carl Barth, the most orthodox of Taylor's lieutenants and the man who led the transformations at both Link-Belt and Tabor, insisted that, "Scientific Management can be used without the employment of any piece-work bonus system" (Nelson 1980, 148). With rare exceptions, as we shall see, those European firms and governments, such as Renault and the Soviet Union, that tried to implement time studies and the differential piece-rate system before they initiated other aspects of Scientific Management ran into great difficulties while those that started with other features were vastly more successful.

While Taylor's version of Scientific Management came to symbolize the entire emphasis on industrial efficiency, there are at least two areas where Taylor's personal vision was much more limited than the broader movement which it symbolized and encouraged. First, Taylor was convinced that no incentives were as effective in motivating workers as differential pay-rates. So, although he was occasionally sympathetic to non-fiscal incentives that singled out individuals for special consideration, he was impatient with employers who engaged in what he called "welfare work," including the provision of such items as lunch rooms, recreational facilities, retirement plans, medical benefits, and insurance (Nelson 1980, 200–201). Second, though Taylor claimed that it was important to match the individual employee to a specific job, he showed no interest in the burgeoning use of intelligence tests and aptitude tests by personnel experts (Nelson 1980, 200).

In spite of Taylor's views and those of his closest associates on such issues, Scientific Management gradually came to symbolize and incorporate broader issues connected with labor reform and personnel management. Indeed it was Lillian Gilbreth, who was at least a sometime member of Taylor's followers, who published the first book to focus on non-financial motivational issues in her *The Psychology of Management* of 1914. There she promoted such techniques as the use of "happiness minutes" to reduce mental fatigue (Haber 1964, 39).

Consistent with the frequent concerns of both labor and owners, Scientific Management focused almost entirely on the efficient production of material goods, measured by the ratio of the exchange value of output—that is, goods produced—to the input costs of labor and materials. It explicitly ignored the distribution of the increased value generated. Indeed, Taylor argued that under Scientific Management such a surplus of goods would be produced, that both owners and workers should, "take their eyes off the division of the surplus as the important matter, and together turn their attention to increasing the size of the surplus" (1911, 27).

The Initial Popularization of Scientific Management

During 1910–1911, Louis Brandeis, a Progressive lawyer, recruited a group of Taylorite experts to testify in public hearings before the Interstate Commerce Commission in a case that would bring Scientific Management—a term which Brandeis popularized to replace systematic management, shop management, and task management, all of which Taylor and his followers were more inclined to use before 1910—to the attention of huge audiences beyond engineers and the owners and managers of factories. Eastern railroads had asked approval for a rate hike that shippers opposed as a form of price gouging. Brandeis, representing the shippers, brought forward his series of witnesses to demonstrate that by increasing railroad efficiency the roads could increase profits by nearly a million dollars a day without the requested rate hike. This case received extended coverage in all major American newspapers. It served to popularize Scientific Management and it gave Taylor the impetus to publish *The Principles of Scientific Management,* which he had written with Morris Cooke's help between 1907 and 1909 (Wrege & Stotka 1978, 736–749). This short and simple work reached huge audiences in both the United States and beyond. Opposition from within the Society of Mechanical Engineers prohibited their sponsorship of *The Principles of Scientific Management,* but the editor of *American Magazine*, a popular reform-oriented publication, invited Taylor to publish it serially. Taylor accepted the invitation because it would offer maximum public exposure of his ideas, but he also had the book privately printed to distribute to management engineers (Schachter 1989, 52).

Except for a description of Frank Gilbreth's time and motion studies of brick laying, which replaced the discussion of Thompson's methods of time studies, there was virtually no content in *The Principles of Scientific Management* that had not been discussed in *Shop Management* eight years earlier. The presentation, however, was revised to appeal to a much broader audience. For example, the earlier relatively bland discussion of how Taylor raised the productivity of pig iron loaders at Bethlehem steel from 1903, was transformed into the unforgettable, inaccurate, and disturbing story of how "Schmidt," a strong and stupid German laborer was effectively tricked into quadrupling the amount of pig iron he loaded each day for a 35 percent increase in pay.

Henry Noll, the man after whom Schmidt was modeled, was an ambitious worker who one historian has identified as "a working class version of Frederick W. Taylor" (Nelson 1980, 172). There is no evidence that he was of less than average intelligence. He was literate and had attended school until age eleven. He served as a volunteer fireman, and he built his own house after working hours (Kanigel 1997, 317). However, that is certainly not how Taylor portrayed him in the story which came to symbolize the insensitivity and brutality of Scientific Management to many workers in the United States and Europe.

Taylor began his story by illustrating the principle that the right kind of employee had to be chosen for each job. In the case of pig iron handling he wrote that a good worker should be "so stupid and so phlegmatic that he more nearly resembles in his mental makeup the ox than any other type." So Taylor described Schmidt as "a man of the mentally sluggish type . . ., a man so stupid that he was unfitted to do most kinds of laboring work" (Taylor 1911).Next, he illustrated the use of higher wages as a motivational strategy. Identifying Schmidt as so focused on money that "a penny looks about the size of a cartwheel" to him, Taylor offered the following imaginary dialog:

> "Schmidt, are you a high-priced man?"
>
> He said, "I don't know what you mean."
>
> I said, "of course you know what I mean. Are you a high-priced man?"
>
> "I don't know what you mean."
>
> I said, "Don't trifle with me. Of course you know what I mean. I want to know whether you are really a high-priced man, or one of those cheap workers satisfied with $1.15 a day," the wages paid up there at that time. . . . "I am looking for fellows I can pay $1.85 to." [the actual wages were only increased to $1.70 per day]. . . .
>
> "I want to find out if you are a high-priced man. If you are, see that pile of pig iron. [47 tons rather than the 12.5 tons moved by the average worker] See that car. If you are a high-priced man you can load that iron on that [car] for $1.85 a day."

The greedy Schmidt, unaware of how much more he is being asked to do, agrees to try to be a high-priced man for the extra $.70 per day; but Taylor has one more demand for Schmidt: he insists that Schmidt follow exactly the sequence of motions and rests designated by the engineering manager:

> "Do see that fellow there?"
>
> "Yes"
>
> "A high-priced man does just as he is told and don't go jawing back . . . When he tells you to pick up a pig you pick it up and walk with it. When he tells you to sit down and rest, you sit down and rest. That is the difference between a high-priced man and those fellows that get $1.15 a day."

Schmidt actually accomplishes the goal set for him, and Taylor concludes the discussion of how Schmidt was motivated by saying—quite incorrectly—

that Schmidt, "had no notion that he was handling four times as much as before. We did not propose to let him know it (Taylor 1911, 41–47)."

Finally, Taylor discussed how the training and financial success of one worker—Schmidt—could be used to transform the productivity of a whole class of workers. While admitting that four out of five workers quit because they could not or would not load 47 tons per day, Taylor wrote that eventually 140 men were doing the job previously done by 600, and he later bragged that those who remained, "constituted the finest body of picked laborers that the writer has ever seen together" (Kanigel 1997, 323).

Ironically, though Taylor came to symbolize the barbarity and insensitivity of Scientific Management in large part as a result of his treatment of Schmidt in *The Principles of Scientific Management*, it seems that he was personally well liked by the workmen at Midvale Steel and probably at Bethlehem as well (Nelson 1980, 39), largely because he followed advice which he offered in one of his earliest technical papers:

> The employer who . . . either talks to his men in a condescending or patronizing way, or else not at all, has no chance of ascertaining their real thoughts or feelings. . . .
>
> Above all it is desirable that men should be talked to on their own level by those who are over them. Each man should be encouraged to discuss any trouble which he may have . . . with those over him. . . .
>
> It is not the large charities . . . so much as the small acts of personal kindness and sympathy, which establish a bond of friendly feeling between them and their employers. (Nelson 1980, 46)

Those who chose to model their treatment of laborers on the pattern set by Taylor's Schmidt story usually failed in their attempts effectively to implement the practices associated with Scientific Management, while those who chose to do what Taylor did, rather than what he said in *The Principles of Scientific Management* were often more successful.

In spite of his condescending and patronizing portrayal of Schmidt, Taylor concluded *The Principles of Scientific Management* with a claim that the whole people, "will no longer tolerate the type of employer who has his eye on dividends alone, who refuses to do his full share of the work, and who merely cracks the whip over the heads of his workmen and attempts to drive them into harder work for low pay." Nor, he claimed, would they, "tolerate tyranny on the part of labor, which demands one increase after another in pay and shorter hours while at the same time it becomes less instead of more efficient." Scientific Management could solve both problems, doubling the productivity of both men and institutions while increasing both wages and profits and offering to everyone, "increased opportunities for education, culture, and recreation" (Taylor 1911, 142).

Early Extensions of Scientific Management Beyond the Shop Floor

By 1911, Taylor had convinced himself and many other engineers that the various practices associated with their attempts to improve efficiency could be associated with two basic claims—first, he claimed that they would create "a complete mental revolution" on the part of workers and owners alike (1911, 26), and second, he insisted that his practices could establish objectively "the one best way" to accomplish any task. Furthermore, he argued that Scientific Management could and should be applied beyond the shop floor and factory. According to Taylor, "the same principles can be applied with equal force to all social activities: to the management of our homes; the management of our farms; the management of the business of our tradesmen, large and small; of our churches, our philanthropic institutions, our universities and our government departments" (1911, 8).

Within a few years of the conclusion of the Eastern Rate Case and the publication of *The Principles of Scientific Management*, Taylor was speaking to enthusiastic public crowds as large as 69,000 at the National Efficiency Exposition and Conference in New York (Haber 1964, 61) rather than the few score engineers that he had addressed previously; his views began to appear in such popular venues as *Atlantic Monthly* and *Ladies Home Journal*; and his ideas were being taught at such places as the Harvard Business School. Furthermore, within a few years *The Principles of Scientific Management* was translated into Chinese, Dutch, French, German, Italian, Japanese, Lettish, Russian, Spanish, and Swedish; and everywhere it promoted Taylor's expansive vision of Scientific Management.

Samuel Haber has demonstrated in *Efficiency and Uplift: Scientific Management in the Progressive Era* (1964) that Taylorism became central to the efficiency fads in personal character, household management, farm management, library management, church management and educational administration, as well as in businesses of all types from 1911 into the 1920s in the United States (Haber 1964, 62–73). Furthermore, as we shall see in the next chapter, Scientific Management also became absorbed by the "Public Administration "movement which had been initiated in 1887 by Woodrow Wilson in "The Study of Administration" but given its major impetus around 1900 by Frank J. Goodnow in *Politics and Administration: A Study in Government* (1900). Goodnow's work, which borrowed heavily from Scientific Management, gave rise to what Dwight Waldo labeled "The Administrative State" (Waldo 1948 & 1984). Though initially much more important at the local level, government through expert administration has gown increasingly important within both national and international contexts since WWII. Thus Taylor's hopes for the extensive public application of Scientific Management were eventually realized.

What Taylor barely understood, even after he had been consulted by scholars and businessmen from all over Europe and Asia, and after *The Principles of Scientific Management* had been translated into at least ten languages, was that Scientific Management would not only become a central feature of Progressive politics in America, but that it would also become a global phenomenon not only in capitalist, but also in socialist and fascist societies. Those mid-twentieth-century scholars who saw American and Soviet societies converging toward one another during the Cold War and long before the advent of *Peristroika,* did so largely because both societies embraced the spirit, if not all of the details, of Scientific Management in shaping both productive industries and governmental policies and procedures (Anderson, 1985).

Edwin Layton has cogently argued that the Scientific Management movement was initially part of a broader reform movement among American engineers intended to give engineers greater autonomy and prestige by undermining their perceived subordination to business interests. One general way to accomplish this was to identify with the ethical codes of physicians and lawyers, which demanded the right to independent judgment so that they might serve the public good. According to Frederick Newell, who was president of the American Association of Engineers from 1919 to 1923, the professional engineer should be "working independently, directing his own affairs with the maximum of personal freedom," much as the doctor or lawyer, who "is called *not* to carry out the instructions of an employer, but, on the contrary, to dictate to the man who ultimately pays the fee . . . he ceases to be a professional man the moment he takes orders from an employer" (Layton 1986, 118). Similarly, Taylor and his colleagues insisted on independence from their employers' views so they might initiate rational reforms in practice, even when such reforms might reduce the influence of owners or fail to increase the profitability of their businesses. Taylor testified before a Congressional committee that it was "inconceivable that a man should devote his time and life to [Scientific Management] for the sake of making money for a whole lot of manufacturers" (Haber 1964, 29). Carl Barth argued that the system should be adopted even when it was "not a paying investment" (Haber 1964, 17). And according to Morris L. Cooke, the engineer is responsible "for the development of facts, regardless of whose advantage they serve" (Jordan 1994, 38).

The general principle that it was the obligation of professionals to serve the public good rather than simply the interests of their employers when the two came into conflict tied Scientific Management to Progressive politics in America and, as we shall see, made it palatable to many others beyond the United States who were deeply critical of capitalism. One of the central claims of Progressive politicians was that unrestrained capitalism was exploiting natural resources and undermining the efforts of small businesses to the detriment of all. This central claim was at the heart of the conservation movement, which Gifford Pinchot expressed in the following way: "Conservation means the greatest good to the greatest number, for the longest time." thus, "Natural resources must be developed and preserved for the benefit of the many, and not merely for the profit of the few" (1910, 46–47). Furthermore, it was at the heart of Theodore Roosevelt's attempts to curb the power of monopolies like Standard Oil. Henry Gantt, Taylor's closest associate, mirrored this view, complaining that the captains of industry were more interested in "harvesting dollars" than in producing the goods that would improve people's lives (Jordan 1994, 59).

A second central claim of Progressive politicians was that the public good would be served by focusing on issues of efficiency. Governmental efficiency and the elimination of corruption was a major concern, but inefficiencies in business were also important. As early as 1895, Louis Brandeis was linking productive efficiency with the public good. "Efficiency means greater production with less effort and at less cost, through the elimination of unnecessary waste, human and material. How else can we hope to attain our social ideals?" (Brandeis 1965, 51)

Such views not only linked Taylorite Scientific Management with Progressive politics in America, they made the *Principles of Scientific Management* one of the most important documents of the Progressive era.

Chapter Two

Scientific Management Enters the Public Domain

The Birth of the Administrative State and the Rise of Modernism in the Arts

In 1948, Dwight Waldo published *The Administrative State*, which was and remains the most widely read and appreciated history of the discipline of public administration as it was developed prior to WWII. Waldo viewed Taylor's work both as the most important theoretical influence on the discipline (1984, xx) and as primarily responsible for four of the greatest weaknesses of the field. According to Waldo those weaknesses included (1) a tendency to assume parallels between business concepts and public sector considerations without adequate analysis of the differences, (2) a tendency to view efficiency as *the* top priority for public administrators rather than as a technique for meeting public demands, (3) a tendency to respect facts over values, and (4) a tendency to ignore the importance of workplace democracy (Schachter 1989, 15). Waldo's views remain dominant among students of public administration, which would be of relatively little interest from our current perspective if it was not the case that the perspectives of public administrators have played a huge role in shaping government entities, not only in the United States, but throughout the world during the twentieth century.

In this chapter we will explore the close connections between Scientific Management; the early ideas of the formulators of public administration; and the changes in actual forms and practices of government agencies, especially at the local level, during the early twentieth century in the United States. In the process, we will suggest why, though Waldo's analysis may sometimes fit the practices associated with early Scientific Management, they are probably less justified as criticisms of both the theoretical orientation of Scientific Management as applied to public entities and as criticisms of later twentieth-century public administration. In addition, we will explore the ways in which Modernism, as a movement in the arts and literature, was influenced by elements associated with Scientific Management.

Scientific Management in Government Industrial Settings

In the United States, the movement of Scientific Management from the private domain into the public domain had already been initiated around 1906. Holden A. Evans, who was in charge of the navy's Mare Island shipyard in San Francisco from 1906 to 1908, had not only seen Taylorism in practice, but he had read *Shop Management* and, equally importantly, he discovered that his chief mechanic had worked at the Tabor Company. Between the two of them, they made Mare Island the first scientifically managed governmental institution:

> Evans first introduced Taylor's refinements of systematic management. He developed a more detailed cost system, reorganized the machine tools, supplied the men with additional tools, and installed Taylor's belt maintenance methods. He also established a central tool grinding department and improved the operation of the tool rooms. . . . He set up a planning department to control production in the machine shop and later, in other areas of the plant. His planning office operated much like the ones Barth had installed at Link-Belt and Tabor, though with fewer clerks. Evans also created a full complement of functional foremen except for the disciplinarian. (Nelson 1980, 156)

Only after all this did Evans initiate time studies and the differential piece-rate system for civilian workers, and by then there was little opposition from workers, although their union leaders sought to have Evans replaced (Nelson 1980, 157).

Several officers associated with the army's Watertown Arsenal had served at Midvale or Bethlehem Steel as inspectors, and a Taylor-designed grinding machine had been in place since the 1890s; so in spite of the fact that union opposition to Scientific Management was more effective at Watertown than at Mare Island, General William Crozier, a Taylor follower, began to implement elements of Scientific Management after the 1908 election under pressure from a number of subordinate officers. Carl Barth was hired in June of 1909 to implement Scientific Management at Watertown. Once again, improvements in machinery, belting, tools, routing, and so forth, were undertaken first. But two of Crozier's officers, Lieutenant Colonel Charles Wheeler and Major Clarence Williams, became impatient and insisted on instituting their own incentive system and initiating time studies before adequate preparation had been done. Taylor warned General Crozier against taking shortcuts, but time studies began in June of 1911 and foundry workers went on strike (Schachter 1989, 61–62).

Eventually union opposition pushed Congress to initiate an investigation of the application of Scientific Management at the arsenal by a three-person sub-committee of the House Labor Committee. Taylor's testimony to the sub-committee suggested that it was the arsenal's failure to follow his and Barth's guidelines that created the great tension at Watertown. It allowed Taylor to emphasize that his goal was and always had been to harmonize the interests of managers and workers to allow, "the substitution of hearty brotherly cooperation for contention and strife; of both pulling hard in the same direction instead of pulling apart" (Schachter 1989, 64). While the Congressional sub-committee disappointed the unions by failing to condemn Scientific Management, one of the most important outcomes of the investigation was to point out that though Taylor and some of his immediate followers might be idealists with pure motives, there was no effective way to stop imitators and others from using elements of Taylorism—especially the time study techniques for setting productivity standards—as a club to force workers to work unreasonably fast without adequate compensation. Morris Cooke admitted as much when he made a special point of arguing that, "there is nothing to prevent scientific management from becoming a nuisance—even a menace—. . . But the same criticism can be made of democracy—even of Christianity—if either is allowed to drift from a genuine purpose or into the hands of charlatans" (Cooke 1913, 493).

Unfortunately no person or group of private persons can stop those who follow from appropriating ideas selectively to serve their own needs. On the other hand, government regulation can sometimes help. To limit the opportunities for using Taylor's methods to exploit labor, Congress passed the Dietrick Amendment to the Army and Navy Appropriations Bills in 1914. This amendment insisted that no funds be allocated to time studies (Shachter 1989, 66); and the Congressional restriction was maintained into the 1950s. One might argue that prohibiting time studies was an unnecessarily blunt instrument for limiting their exploitative use, but this is often the nature of regulatory legislation. In the meantime, other elements of Scientific Management implemented at Watertown led to dramatic increases in productivity, and elements of Scientific Management spread to many other arsenals.

Both navy and army experiences between 1906 and 1911 established that a relatively flexible approach to Scientific Management could be effective in increasing productivity at government factory-like installations even when one of Taylor's favorite elements could not be implemented because of union and Congressional opposition. So when WWI began, General Crozier, as head of Army Ordinance, hired every member of the Taylor Society that he could find (Kanigel 1997, 487)—and most of those that he could not get were already committed, like Gantt, to the navy or to President Wilson's Council of National Defense.

When the Taylor Society was formed by members of the pre-existing Society to Promote the Science of Management after Taylor's death in 1915, one of its major goals was to oppose misapplications of Scientific Management, and, "to preserve, in the midst of a rush of industry to get efficient quickly, the engineering technique and the idealism of what had been given the distinguishing name of Scientific Management" (Schacher 1989, 69); There was, however, little that the Taylorites could do to stop being identified with their less than idealistic imitators.

Initial Non-industrial Applications of Scientific Management

After the Eastern rate case, most early advocates of Scientific Management continued to focus on industrial applications. A few, however, including both Frank and Lillian Gilbreth, Henry Gantt, Taylor, and Morris Cooke turned their attention to spreading their principles into broader domains. Frank Gilbreth, for example, expanded his early time and motion studies connected with brick-laying and factory tasks to the work of medical personnel, including surgeons in his *Motion Study* of 1911; and he and Lillian sought to help the handicapped through their *Motion Study for the Handicapped* of 1920. Taylor spent much of his time between 1911 and his death in 1915 introducing Scientific Management into business schools, including those at Harvard, Columbia, and Dartmouth. More importantly, many members of the Taylor Society and its precursor, the Society to Promote the Science of Management initiated the teaching of Scientific Management as part of broader engineering and business programs throughout the country, transforming business education from accounting-based to management-based programs (Nelson 1992, 77–101).

At the same time Taylor also began to think about applications to non-industrial governmental issues. One of his last papers, "On Government Efficiency," promoted Progressive efforts to replace the spoils system of political appointments with a system based on relevant expertise (Schachter 1989, 73) and suggested a cabinet level efficiency expert to oversee the implementation of Scientific Management in all departments, beginning with "the scientific study of what should constitute a proper days work for each government employee." (Haber 1964, 107) Unfortunately Taylor did not suggest how efficiency should be measured outside the realm of material production.

An Example from Education

Among the first non-industrial and non-business enterprises to respond to the appeal of Scientific Management was American public education. At its annual meeting in February of 1911, the Department of Superintendence of the National Education Association (NEA) heard J. George Becht, principal of the State Normal School in Claxton, Pennsylvania, give a speech urging educators to embrace the pioneering efforts of Taylor. That summer the NEA established a committee on "Tests and Standards of Efficiency of Schools and School Systems," and in the fall the High School Teachers Association of New York began a three-year study of "Efficiency in the High Schools through the Application of the Principles of Scientific Management" (Callahan 1962, 54–55). This group hired one of Taylor's competitors, Harrington Emerson, to help in its efforts, and Emerson focused on standardization, insisting on the need for "Standard records, Planning, Standard conditions, Standardized operations, Standard Instructions, Standard Schedules, and Efficiency rewards" (Callahan 1962, 56).

Though the consequences of this committee's actions were meager, the NEA's fascination with Scientific Management grew over the next several decades. One consequence was that schools of education began to focus on scientific research on the effectiveness of different educational procedures, using standardized tests, surveys, efficiency rating systems, and elaborate systems of records and reports, giving "at least the impression of providing a 'factual' basis for education" (Callahan 1962, 67). A second consequence was that between about 1912 and 1930 in most cities the primary responsibility for the character of primary and secondary education was taken out of the hands of the general public and placed in the hands of special educational experts trained in educational administration.

Previously, in most cities each ward had its own elected board of education which could promote education suited to the needs of the local population, whatever its ethnicity and occupational makeup. But such education was now seen as inefficient. Reformers sought to centralize and standardize educational administration, put policy in the hands of a small school board made up largely of businessmen advised by an expert school superintendent, and they viewed educational management as very much like the management of a manufacturing enterprise. In 1916, for example,

> Professor Franklin Bobbitt of the University of Chicago summarized "The principles of good management" in two columns, one for a manufacturing company employing 1,200 and the other for a school system of the same size. . . At the end Bobbit concluded that "when it was asserted that educational management must in its general guidelines be different from good business management, it can be shown from such a parallel study that there is absolutely no validity to the contention". . . . In a survey of San Francisco schools, the

> U.S. Commissioner of Education, reproduced Bobbit's chart verbatim. (Tyack 1974, 144)

Like the scientific managers of industrial enterprises, the new educational managers were relatively uninterested in the distribution of advantages produced by their more efficient administrations. In particular they saw the modification of education to meet the needs of special groups, including ethnic minorities, people with handicaps, and so forth, as both inefficient and unnecessary, setting a tone for American urban education that lasted into the civil rights era of the 1960s.

Finally, for practical purposes, the dominant measure of educational efficiency and effectiveness, like the measure of industrial efficiency, became the cost of instruction per pupil, or "investment per pupil" (Callahan 1962, 70–71). For one of the most widely read theorists of educational management, Frank Spaulding, even the value of different subjects in the curriculum should be based on the relative hourly cost of instruction per student in that subject. Thus, because 5.9 pupil recitations in Greek cost the same one dollar as 23.8 pupil recitations in French, "we ought to purchase no more Greek instruction at the rate of 5.9 pupil recitation per dollar. . . The price must go down, or we shall invest in something else" (Callahan 1962, 73). While there may have been legitimate pedagogical and sociological reasons for removing Greek from American high school curricula it does seem that economic issues were central to the choice. From this perspective, one of the most effective ways of increasing efficiency was to simply increase class size, which is what Spaulding did as superintendent of schools in Newton, Massachusetts, regardless of any consideration of the impact on teachers or students.

Henry Gantt and the "New Machine"

Among Taylor's close associates, Henry Gantt and Morris Cooke became the two greatest promoters of Scientific Management in government. Of these two, Cooke had the longest term impact, but Gantt was far and away the most radical and aggressive. Gantt had read Thorstein Veblen's earliest works, which condemned the exclusive emphasis of capitalism on "pecuniary" motives, and argued that engineers, with their expertise and focus on "facts" as well as the public interest should be in charge of American industry. But Gantt carried Veblen's criticisms even farther into a Leninist reinterpretation of political democracy. According to Gantt, "true democracy is attained only when men are endowed with authority in proportion to their ability to use it efficiently and their willingness to promote the public good" (Haber 1964, 48). In another speech, Henry Gantt defined the "new democracy," saying that it "does not consist in the privilege of doing as one pleases, whether it is right or wrong, but in each man's doing his part in the best way that can be devised from scientific knowledge and experience" (In Maier 1970, 33). Explicitly rejecting the "debating society theory of government," Gantt argued that government should be based on facts rather than opinions and that, "the engineer, who is a man of few opinions and many facts and many deeds, should be accorded the economic leadership which is his proper place in our economic system" (Haber 1964, 48).

In December of 1916 Gantt recruited about fifty members of the Association of Mechanical Engineers to establish The New Machine, which was intended to promote his vision of the new democracy. Gantt and his New Machine colleagues clearly went far beyond where even most reform-minded engineers could follow when they sent Woodrow Wilson a letter asking that he grant them presidential authority to manage American industry—a letter which Wilson pointedly ignored (Haber 1964, 47). Though Gantt was unimpressed with Bolshevik leadership in Russia, he admired the Soviet arrangement in which the industrial system was supposed to be directed at serving the community rather than creating profits for a few owners, and at least one member of the New Machine, Walter Polakov, went to Russia to help the Soviets create their first Five Year Plan (Merkle 1980, 122).

During WWI, the New Machine was disbanded, but Gantt and many of its former members, like other Scientific Management advocates, took up government posts and helped to orchestrate war-time production and the allocation of resources, focusing attention on production rather than profits. Writing in 1919, Gantt argued that the war had demonstrated the inability of democratic politics, without the help of engineers, to cope with modern warfare and suggested that engineers should continue their leading public role in peacetime. But his hopes that the public service orientation of war-time industries would carry on after the war and that leadership would remain in the hands of engineers were soon shattered.

Gantt and the advocates of Howard Scott's Technocracy movement a decade later represented the radical fringe among those connected with Scientific Management. They manifested all of those attitudes that Putnam identified with technocrats, and had their views dominated early public administration, all of Waldo's critiques would have been fully justified.

Morris Cooke, the First Public Administrator

In the same year that the Eastern Rate Case broke, reform mayor Rudolph Blankenberg of Philadelphia tried to hire Taylor as Director of Public Works to both increase the efficiency of city employees and to exert control over the privately owned public utilities serving Philadelphia. Taylor refused, but he recommended his collaborator, Morris Cooke, who claimed that "the engineer above all others is qualified to be responsible to solve the great vital problems of our municipalities" (Jordan 1994, 39). Like Taylor, Cooke was a member of the ASME (American Association of Mechanical Engineers) and had an engineering degree; but he came to Scientific Management in 1903 from a position as an executive in the publishing business, hoping to apply the insights from *Shop Management* to the print shop rather than the factory. Under Taylor's tutiledge Cooke worked at implementing Scientific Management in traditional industrial environments; but his interests were always in extending Taylor's ideas to new settings. From his perspective, the "scientific spirit," as opposed to the "craft spirit" focused attention on the similarities between various kinds of activities, and "Scientific Management asks of the paper-maker, the college professor, the government official, and the iron master that each drop his craft spirit and take on as much as he can of the scientific spirit" (Cooke 1913, 484).

Cooke had, for instance, agreed to head a Carnegie Foundation for the Advancement of Teaching study of the efficiency of colleges and universities in 1909. Taking to heart the notion that it was critical to have some agreed upon measure of efficiency to apply, Cooke devised the "student contact hour" as a measure of educational efficiency for his 1910 report to the foundation, anticipating Spaulding's use of a similar measure for primary and secondary education (Kanigel 1997, 490). Faculty understandably objected, arguing that just as not all workers were created equal, not all contact hours were of the same quality; but when resource allocation to various departments is discussed at almost every college in the nation today, one of the most crucial questions is still which departments have the highest average number of student contact hours per faculty member.

Given his background and attitudes, Cooke welcomed Blankenberg's invitation to shape Scientific Management to issues faced by municipal governments. From that time forward until his death in 1960, Cooke focused his attention on public service and was usually employed by some governmental agency. After serving as head of Philadelphia's Department of Public Works from 1911 to 1915, he served under Franklin Roosevelt as chair of the Public Works Administration's Mississippi Valley Committee, director of the National Resource Board's Water Resources Section, and as chief administrator of the Rural Electrification Administration.

At Philadelphia, in spite of bitter opposition from more conservative engineers and businessmen, Cooke was able to reduce the cost of such city supplied public services as street maintenance up to 50 percent. In order to do so he instituted empirical studies of the most effective ways to pave streets, remove snow, and so forth. Like Taylor, he insisted that city employees learn to undertake their tasks as his planners recommended; but also like Taylor, he then invited employees to recommend improved techniques. After doing experiments to determine which method was more efficient, he then adopted that procedure (Schachter 1989, 78).

Above all, Cooke argued that Scientific Management depended upon establishing facts:

> We do not ask anybody to accept our personal opinions, nor will we accept those of others. Facts must determine each question as it comes up. If you haven't the facts you must get them, whether they can be secured in five minutes at no expense, or whether it takes thirty years and the expenditure of hundreds of thousands of dollars, as it did in the case of the development of slide rules for machine tools. (Cooke 1913, 488)

Cooke had independent organizations determine whether contract stipulations were being met by private service providers to Philadelphia, and when they were not, he arranged to have substantial fines levied against the contractors (Schachter 1989, 84–85). In addition, Cooke undertook a joint effort with several other cities to establish the costs that private utilities incurred in providing services. Though the utilities tried to block the public's access to relevant information, and though he was censured by the ASME for embarrassing some of his fellow engineers for their stonewalling (Schachter 1989, 88), Cooke was eventually able to force a rate reduction by the private utility which supplied electricity to Philadelphia that saved the city $200,000 per year and its citizens over $1,000,000 annually (Layton, 160–63), demonstrating the ability of scientific management to serve the public interest and once more promoting wide spread interest in the movement.

While Cooke's way of implementing Scientific Management in municipal governments was virtually always consistent with Taylor's views, his choice of what to emphasize was often different. For example, Cooke harkened back to an argument made in 1885 by ASME member Henry Metcalf, who had argued that public administrators face dual pressures. Like those who work for private clients, public administrators seek to improve efficiency, but they must also be accountable to legislative demands which may override considerations of efficiency. Indeed, for Cooke, unlike for Gantt and his followers, citizens' groups or their elective representatives should always set public agency agendas (Schachter 1989, 75–76). Cooke did not, however, agree with those public administration advocates among the Progressives who thought that technical administrative work could clearly be separated from political decision making. For Cooke, truly democratic decision making must be based on adequate information, and this information almost always had to be provided by experts.

Cooke's most extended discussion of the inevitable intertwining of administration and politics appeared in two of his early classics dealing with municipal public administration, *How About It* of 1917 and *Our Cities Awake*, published in 1918. Cooke pointed out that in increasing numbers of cases at the beginning of the twentieth century, new scientific discoveries generated new public demands. Thus, for example, Cooke instanced the biologists' discovery that Typhus was a waterborne disease and argued that this discovery had produced a citizen demand for clean water and for monitoring water quality (Cooke 1918, 267). Once this general demand had been created, expertise played a key role in a second phase of public decision. How should the water be purified? Is filtration enough? Is chemical treatment more cost effective? Or should the city spend more money and combine filtration and chemical treatment? In all such cases the chief obligation of the administrative expert is to develop "facts" about various options. Until administrative experts convey information about the costs and effectiveness of different treatment options, citizens cannot make a sensible decision. Moreover, the administrative expert has an obligation to assess and communicate the differences between long-term and short-term efficiencies and to advocate for what she judges to be in the best public interest. Inevitably this situation places the administrator centrally *within* the political process.

Ultimately, according to Cooke, the administrator must accept the public's decision even if she disagrees with it: "If citizens value short term savings to the extent that they will not authorize the expenditures that are necessary to secure the paving or street cleaning that is most efficient in the long run, the agency's obligation is to provide the inefficient services citizens desire" (In Schachter 1989, 76–77). Until that decision is finalized she has a political role to play (Cooke 1917, 17). In acknowledging that citizens' values can—indeed, must—override considerations of fact when they conflict, Cooke, at least, promoted a vision of public administration that did not completely privilege facts over values.

According to Cooke, once the citizens give an agency its marching orders, then the agency's job is to provide the required service at the lowest possible cost. Thus, efficiency becomes a means for responding to public demands rather than an end in itself. In order to meet public demands most expeditiously Cooke accepted Taylor's argument that organizations must hire first-class people, use scientific methods to establish work processes, and motivate their people to work effectively. Like Taylor, Cooke preferred to choose personnel based on specific job-related criteria rather than on the basis of field-specific but very general written civil service exams. His compromise proposal was to let agency heads choose from the top four test takers based on more narrowly focused interviews (Cooke 1918, 167).

Cooke's discussions of how to motivate workers were more detailed and expansive than that of other Taylorites because governmental regulations did not allow for pay-rate differentials. He urged close relations between supervisors and subordinates because, "Nine times out of ten you will find a sick child or wife at home, the worker in debt, or some other purely personal reason for work below par. Working without the closest spirit of confidence and cooperation, no opportunity is afforded for finding out these sources of inefficiency" (Cooke 1913, 491). In addition, he argued that morale was greater where collaborative decision making was practiced; so he promoted workplace democracy. Furthermore, he argued that, "functional management leads ultimately to the democratization of our industries, [because]. . .the shop is run by the collective intelligence of the many, rather than the 'I say so' of those who, in the language of the day, own the business" (Cooke 1913, 491). If directions had to be given from top down, he insisted that leaders could motivate employees by never giving orders that could not be justified to those who would carry them out.

Most importantly, Cooke emphasized the notion that employees should be offered training which would allow them to move into higher level and more challenging work so that, "the worker is inspired with joy in his task" (Schachter 1989, 82). If there was no path to more rewarding work in the worker's present environment, Cooke even told the management expert that he, "must make it one of your most important duties to see to it that they get this chance for advancement somewhere else" (Cooke 1913, 492). Harkening back to such nineteenth-century technocratic theorists as Saint- Simon and forward to the "self-actualization" theories of Abraham Maslow, Cooke thus argued both that task involvement could provide its own satisfaction and motivation and that it was the scientific manager's job to see that each worker found a position suited to her own abilities and psychological needs.

Sometime before any of those directly associated with Taylor's version of Scientific Management became focused on public service, a number of reform-oriented political thinkers—mostly associated with the American Economic Association (AEA) founded in 1885 and the American Political Science Association (APSA), established in 1903 through the fusion of several previously existing regional Political Science Associations—began to emphasize the applicability of the ASME's approaches to systematic management to public entities. As early as 1896, for example, the AEA and the Midwestern Political Science Association had collaborated in reprinting several of the ASME management papers, including Taylor's initial piece-rate article, which had just appeared in the previous year.

On one critical issue—the relationship between administrators and "politics"—however, most of these progressive reformers followed Woodrow Wilson, whose "The Study of Administration" of 1887 was the first formal discussion of public administration in the United States (Wilson 1887, 197–222). Wilson insisted that administration should be separated from politics:

> Administration lies outside the proper sphere of politics. Administrative questions are not political questions. Although politics sets the tasks for administration, it should not be suffered to manipulate its offices. . . . The broad plans of

> governmental action are not administrative; the detailed execution of such plans is administrative. (Wilson 1887, 200, 202)

What distinguishes administration from politics according to Wilson is that administration could be grounded in a science which could, "discover, first, what government can properly and successfully do, and second, how it can do these proper things with the utmost efficiency and at the least possible cost either of money or energy" (Wilson 1887, 197).

Though Wilson's essay seems not to have been widely read, its fundamental distinction was articulated again by Frank Goodnow in *Politics and Administration: A Study in Government* (1900) which *was* widely appealed to by reformers. In line with his views about the separation of administration from politics, Goodnow played a major role in promoting the "city manager" form of municipal government, in which an elected council and mayor presumably determined broad policy while an administrative manager simply carried out its wishes (Haber 1964, 103).

The New York Bureau of Municipal Research: The Institutionalization of Public Administration and Attacks on the "Political" Role of Public Administrators

In terms of the institutionalization of public administration, the creation of the New York Bureau of Municipal Research (NYBMR) in 1907 was especially important because it trained a cadre of public administration professionals and created new institutional settings in which they could work. In 1906, William Allen, a member of New York City's Association for Improving the Condition of the Poor, managed to get funding from several New York philanthropists to transform New York City's Bureau of City Betterment into the Bureau of Municipal Research. Led by Allen; Frederick Cleveland, professor of finance at New York University; Henry Bruere, who had been head of the Bureau of City Betterment; and by Charles A. Beard, professor of Law and Government at Columbia, the ultimate goal of the Bureau of Municipal Research was to create social betterment. That ultimate goal was to be achieved by creating efficient government, which in turn depended on an informed citizenship which could push government agents to act ethically and efficiently. Finally, an informed and efficient citizenry would be achieved by promoting, "the application of scientific principles to government" (McDonald 2010, 818).

Of the initial leaders of the Bureau of Municipal Research, Cleveland was most directly linked to Taylor. Having been put in contact with the Scientific Management leader by Morris Cooke, Cleveland joined the Taylor Society. In addition, Cleveland and Cooke carried on a correspondence about municipal government through the 1920s (Schachter 1989, 92). The rest of the leadership gained their initial commitment to improving government performance through fact gathering and analysis in part through reading Taylor's *Shop Management* and in part through the scientific approach to political issues promoted by the positivistic premises of both the AEA and APSA. In addition to their emphasis on empirical studies, Bureau leaders drew their emphases on functional management, standardization of job descriptions and special planning departments from Taylor's works.

At first the Bureau's efforts to study public work processes and how they might be improved were stymied by political bosses, especially by Manhattan Borough President John Ahern, who refused Bureau employees access to public meetings and records and even denied them entrance to City Hall. Doing an end run around government officials, Bureau employees did a study of city street repair in Manhattan which included such details as comparing their count of bags of cement used with the inflated number claimed in the publicly available budget. The results, reported in 1907 as *How Manhattan Is Governed* embarrassed Ahern, who claimed ignorance of the corruption uncovered and who sued the Bureau for libel. The public scandal, however, not only caused Ahern's resignation, but it also forced Bureau access to virtually all agencies' records (McDonald 2010, 819). According to New York state's reform governor Hughes, who requested Ahern's resignation as a result of the Bureau study, "The character of the bureau's investigation to aid administration in city affairs marks one of the most important improvements of recent years. It is striving to get at the facts in an honorable, straight forward way, and is striving to present them so that they will be intelligently comprehended" (Hopkins 1912, 239).

Undoubtedly the most important result of the splash made by *How Manhattan Is Governed* was the creation of their own bureaus of municipal research by cities and citizens' groups around the country, starting with Cincinnati, Chicago, Cleveland, Memphis, Philadelphia, and Saint Louis, and spreading to such small "mountain villages" as Dobbs Ferry, New York (Hopkins 1912, 237). Between 1911 and 1917, sixteen states also created Efficiency Commissions, and in 1910 President William Howard Taft created the United States Commission on Economy and Efficiency. He placed Frederick Cleveland at its head and included Frank Goodnow as a member, bringing Taylorite studies to the federal level where the major emphasis was on planning (Haber 1964, 113). While complicated political reasons led Woodrow Wilson to abandon the Commission in 1913, some of its recommendations have had a huge impact on the federal government. The Commission, for example, recommended the establishment of a budget planning bureau, which, as the Bureau of Management and Budget, provided the leadership in formulating and monitoring the federal budget through most of the twentieth century.

The burgeoning of bureaus and commissions focused on governmental research in turn created a demand for staff to run these organizations. Since the employees of the New York Bureau were the only people with relevant experience, this situation produced a drain on Bureau staff. In order to train staff both for their own organization and for other public service organizations, the New York Bureau, with the initial help of $80,000 from the widow of E. H. Harriman, founded the Training School for Public Service with the following goals:

> To train men for the study and administration of public business.
> To qualify men [sic] to meet the growing need for student and administrators competent: (a) to test and (b) to improve methods and results of public service.
> To publish facts which may be incorporated in textbooks and lectures, in teaching the relation to the public service of: political science, government, and sociology; (b) accountancy; (c) engineering; (d) law; (e) public hygiene; (f) school administration; (g) journalism; (h) medicine, etc. (McDonald 2010, 826).

For the first-year class of 25, there were 485 applicants (McDonald 2010, 819).

The Bureau's Training School demanded that students gain practical experience in doing survey work by pairing them with Bureau staff while they completed a curriculum that began with budgeting practices and expanded to include municipal politics, law, and accounting. In 1913 a new symposium requirement was established. Frederick W. Taylor was one of the guest speakers at symposia. Finally, in 1915, when Beard became director, he added a pre-practice year of courses on "public administration, management, and organization methods" (McDonald 2010, 820). Before 1911 single courses in public administration had been taught at Columbia and Harvard, but the Bureau's Training School provided a model program comparable in many ways to university programs in a wide range of disciplines. Immediately after the school began, imitator programs cropped up throughout the country—at Michigan in 1912, at the University of California in 1915, at Missouri in 1917, at Syracuse University, where it was designated The Maxwell School of Citizenship and Public Affairs, in 1924, and at the Universities of Kansas and Minnesota in 1925 (McDonald 2010, 827).

The initial emphasis of the Bureau of Municipal Research had been on providing information to the broader public on the assumption that an informed public would insist upon effective and efficient government. Even when praising the city manager form of government, which generally assumed a radical distinction between political and administrative functions, William Allen emphasized the informational and thus political role of the managers, who he praised for, "taking the public gently by the hand and helping it to really understand what it wants" (Haber 1964, 104n12). Secondarily, the Bureau offered training to government employees, especially regarding new procedures and standardized systems for accounting and budgeting (Hopkins 1913, 239).

Two different circumstances moved the Bureau's emphasis away from promoting citizen knowledge and toward promoting administrative decisions by professional experts on the public's behalf. The first circumstance reflected the results of a field study done in 1912–13 under the direction of Henry Bruere on the effectiveness of government in ten of the nearly 225 "commission cities,"—that is, cities who turned over many functions including zoning, architectural review, and so forth, to commissions made up of persons with expertise in relevant disciplines, largely taking decisions out of the hands of the public or their elected representatives (Haber 1964, 102n7). Results showed that not only was the physical condition of these cities superior to that of cities without a commission structure, but that citizen satisfaction with governance, the quality and satisfaction of government personnel, and both public records and work methods were all rated highly in commission cities. Thus it seemed that at least some municipal decisions were better made by experts than democratically. As Bruere wrote: "Citizens of larger cities must frankly recognize the need for professional service on behalf of citizen interests. . . . Even efficient private citizens cannot deal helpfully with expert governmental questions. Efficient citizens will evidence their efficiency by supporting constructive efforts for governmental betterment" (Haber 1964, 112).

Much more important in shifting the emphasis of the Bureau was the fallout from the Bureau's positions regarding educational reform initiated by the New York legislature in 1911. The legislature pushed responsibility for local schools on to local Boards of Education, which frequently delegated many of their functions to specially trained and appointed superintendents of public instruction. In New York City, the Board of Estimate and Apportionment was delegated to do a study of city schools and to recommend a governance structure. At least initially the Bureau supported the Board's study; but soon the committee appointed by the Board came into conflict with the Bureau. The committee recommended a small Board of Education independent of the city's elected government that would be heavily dependent on the recommendations of an appointive superintendent of education. Such a structure would minimize political manipulation of education. The Bureau, on the other hand, strongly supported greater citizen involvement in the schools on the grounds that informed citizens would hold professional administrators to higher standards of ethics and effectiveness in educating students.

Members of the committee approached John D. Rockefeller, now the chief financial supporter of the Bureau, complaining of the Bureau's interference and use of publicity to attack their position. Rockefeller responded by having his lawyer, Starr Murphy, inform the Bureau that he believed that educational decisions were best left to professionals and that the Bureau would lose his funding if it did not, "stop its New York school work," and pull back from its emphasis on providing information to the general public (McDonald 2010, 823). Some of the founders of the Bureau, including Allen, resigned rather than accept Rockefeller's restrictions. Others, including Cleveland, remained and focused on doing scientific studies and providing information and recommendations to government officials. Bureau trustees tried to get Murphy and Rockefeller to allow them to go public when the government refused to enact reforms recommended by the Bureau, but Murphy responded by telling them flat out to "stay away from propaganda and focus on the science behind administration" (McDonald 2010, 823).

The upshot of Rockefeller's arguments was, that in spite of its initial emphasis on a political role as the fact collector for and advisor to the general public—a role consistent with the often-repeated views of Cooke and with the broadest understanding of participatory democracy—the New York Bureau of Municipal Research became a reluctant model for those advocates of public administration who sought to separate administration from politics and to focus on management techniques. Henceforth the Bureau would be seen not primarily as an entity created to promote social welfare, using the insights and practices of Scientific Management as techniques for achieving that political goal; but rather as an entity whose primary goal was methodological. In Beard's words, "It may be truly and modestly said that the founding of the Bureau of Municipal Research opened a new period in the methods of research in municipal administration by introducing the methods of natural science, namely the methods of first hand observation" (McDonald 2010, 827).

Under his leadership, education at the Training School became increasingly focused on the "science" of public administration. Initially the only required readings were from Taylor's writings, but as course work increasingly augmented practical experience, instructors at the training school produced most of the early textbooks on public administration from the 1913 *Metz Handbook of Municipal Accounting* through Leonard D. White's comprehensive *Introduction to the Study of Public Administration*, which went through three editions between 1926 and 1948 (McDonald 2010, 828).

"Neutral" Public Administrators and Commission Cities

Those who sought to replace politics with technical decision making like Gantt or who saw administrative experts as inevitably enmeshed in politics, like Cooke and Allen, were vastly less numerous than those who followed the ideas of Goodnow and other non-engineer advocates of public administration, who thought that the appropriate role of technocratic or scientific administration was to determine the most efficient and effective way to implement policies that were arrived at through some form of democratic decision making in places like the United States, or by party leaders or dictators elsewhere. Such views denounced the notion that administrators should be "political" and embroil themselves in public controversies. Instead, their decisions should be based on a scientific knowledge of the consequences of those decisions and their aims should be limited to finding the most efficient way to accomplish goals set by political activity. In line with such views Bruere and Cleveland became strong advocates for the city manager form of municipal government (Haber 1964, 112n36). This form has become the dominant form of governance in most small and moderate-sized cities throughout the United States. At the same time, evidence from studies of the impact of commission structures within municipal governments also suggested that there were significant domains in which public decisions made by experts rather than by elected officials or "debating society democracy" led to better government.

Just where the line should be drawn between questions that should be treated democratically and those made by technically trained experts was—and remains—a matter of substantial disagreement, though trends throughout the world have, as we shall see in later chapters, been to place increasing numbers of decisions in the hands of experts until there is some form of public outcry and resistance.

Herbert Hoover: Consummate Public Administrator

While the establishment of a federal level Commission on Efficiency and Economy and the subsequent creation of the Bureau of Management and Budget signaled the spread of Scientific Management and public administration beyond the municipal government level, the election of Herbert Hoover as president of the United States illustrated the extent to which management engineers and many of their claims to objectivity were admired by the American public during the third decade of the twentieth century. Hoover frequently published in *The Bulletin of the Taylor Society* and wrote the introduction to *Scientific Management After Taylor, 1914—1920*. Furthermore, he had used his management skills during WWI and its immediate aftermath in organizing the delivery of food supplies to Europe. His actions as "Food Czar" in WWI, as Secretary of Commerce, and then as President illustrate both some of the greatest strengths of his version of the public administration mentality and some of its major weaknesses.

Born in 1874, Hoover grew up in a serious and poor Quaker farming family in Iowa. Losing both his parents by age ten, he was taken in by an uncle in Oregon who was a physician and later, a land developer. At thirteen he began to attend the Friends Pacific Academy, where he developed a special interest in mathematics and bookkeeping. When his uncle moved to Salem, Oregon, Herbert kept up his mathematics interest at the Capital Business College; but except in mathematics he felt that his early education was deficient. He worked briefly as an office boy, clerk and then accountant for the Oregon Land Company, and in 1891, he entered the first class admitted to Stanford in spite of his spotty earlier education. At that point Stanford was focused on practical education rather than a broad, liberal education, and Hoover entered as a mechanical engineer, though he subsequently changed to geology to prepare to work in the mining industry in California.

Both his Quaker upbringing and his science and engineering education promoted an "ethos of useful service" in Hoover (Clements 2000, 22). Though as an adult he did not attend Quaker meetings, the early Quaker emphasis on individualism almost certainly promoted his long-term emphasis on voluntarism and his opposition to government coercion in peacetime. Like Morris Cooke, he argued that governmental experts should provide the public with accurate information and advice upon which to base decisions, but that members of the public should have the opportunity to voluntarily comply with that advice or to reject it—except in wartime. Then expertise should rule because, "Democracy is a form of government born of peace, constructed for peace, and maintainable only in peace. To carry on war successfully requires a dictatorship of some kind or other. . . . No other way has ever been found" (Nash 1996, 345). Thus, when Woodrow Wilson tapped Hoover to head the U.S. Food Administration in WWI and Congress failed to authorize him to set prices, he used his authority to license food producers to force compliance with his price recommendations.

His education, coupled with his experience as a mining engineer, also convinced Hoover that advances in technology and efficiency could produce unalloyed benefits for owners and workers alike, creating a better standard of living for all. Speaking of his experiences as a mining engineer he wrote, "Our operations were a demonstration [of] an industrial fundamental—greater technical service, more labor saving devices, lower cost, and larger production *and* higher wages" (Hoover 1951, 89). When Hoover returned to the United States in 1919 after working internationally for twenty-two years, he soon became president of the Federated American Engineering Societies, a position in which he tirelessly campaigned to increase American industrial efficiency and productivity on the assumption that it would benefit everyone. Industrial democracy of the kind promoted by the Taylor Society would surely lead to a reasonable division of the products of industry, and well-paid workers would expand a consumer society that would provide ever growing markets for ever more productive industries.

Though he recognized that increased productivity would lead to a reduced average number of hours needed from American workers, Hoover followed

Taylor's lead in believing that any additional leisure would be widely distributed and constructively used for healthy recreation and for education (Clements 2000, 93). He was incapable of believing that greater efficiency could bring such overproduction and dramatic reductions in employment that they could thoroughly disrupt an economy so that nothing short of a massive government program of public works employment could resuscitate the economic system until it was too late to head off the depths of the Great Depression.

Hoover was also convinced that an engineer's training in the exact sciences "makes for truth and conscience" (Clements 2000, 34), so he was especially inclined to trust engineers and, when possible, chose them as colleagues and subordinates in programs that he was involved with. Since virtually all engineers and technocrats trained in the exact sciences shared his vision of plenty for all through increased efficiency, almost no one that he associated with offered a strong counter to that vision. Finally, Hoover's education convinced him that engineering reformers were right in arguing that engineering should be recognized as a profession comparable to law and medicine and that professional status gave the engineer a special obligation to serve the public rather than to serve only his clients. Like Gantt and Cooke, Hoover was convinced that engineering created men that were independent of any economic or political interests and that engineers "comprise a force in the community absolutely unique in the solution of national problems" (Clements 2000, 36). Thus, after Hoover had worked as a mining engineer and administrator throughout the world—in Australia, Burma, China, and Russia as well as in the United States—and after he had become a partner in the London-based Bewick, Moreing and Company, which developed and managed mines internationally, he decided that he had made enough money and he sought to enter some form of public service.

An opportunity dropped into his lap at the very beginning of WWI. He and his wife Lou, who was also a geology student at Stanford when Herbert attended, were resident in London when the war stranded many American travelers in Europe with no access to their funds and no way to get home. Hoover helped to organize and then directed the Committee of American Residents in London for Assistance of American Travelers, which assisted some 120,000 travelers over a three-month period. As a consequence of his organization of this effort, Hoover came to the attention of the U.S. State Department, which sanctioned him to establish and again manage the Commission for the Relief of Belgium (CRB) during the winter of 1914–1915 to avert the starvation of residents of Belgium and Northern France.

Over a period of two and one-half years the CRB fed approximately nine million people. Hoover oversaw the purchase of food in the United States, its transport across the United States and the Atlantic, and then its distribution through Belgium and Northern France. In the process he hired a small group of paid organizers (mostly engineers with business backgrounds), coordinated some 130,000 volunteers, negotiated with multiple governments to clear obstacles, secured funding, and kept administrative costs below 1 percent of total expenses (Clements 2000, 29–30). The amazing feats of the CRB in turn led Woodrow Wilson to appoint Hoover to head the U.S. Food Administration which was created in 1917 to control the production, sales, consumption, storage, export and import, as well as the transport of food and fuel during the war. His success in running this program without a hint of scandal, again led Wilson to appoint Hoover in 1918 as head of the American Relief Organization to help restore the European economy. As in his other posts, Hoover oversaw a centralized planning effort with disbursed execution by experts.

When Warren Harding was elected to the presidency in 1920, Hoover's reputation as an outstanding administrator led Harding to offer him the position of Secretary of Commerce. In spite of some Republican resistance to the appointment of a person who had never before even voted in a presidential election and whose credentials as a Republican were problematic because of his major roles in the administration of Woodrow Wilson, Hoover was confirmed. He accepted the post only on the condition that he be authorized to undertake, "a vigorous policy of leadership in stimulation of industries to cooperation among themselves in the direction of greater reduction of waste in our manufacturing and distribution methods, in the direction of greater standardization in the production of certain essentials and in the direction of stimulating and organizing our foreign markets." Furthermore, he would be authorized to collect information about society and industry from individuals and organizations (Clements 2000, 59–60). That is, he would be able to use the Commerce Department to forward the agenda of the Scientific Management movement and he would be able to collect the statistical information that would make it possible for him to provide information and advice to businesses that would allow them to voluntarily engage in planning for maximum efficiency in production.

In some domains the Commerce Department under Hoover was successful and its accomplishments were widely admired. The Bureau of Standards played a major role in standardizing many processes and products—that is in standardizing lumber sizes and grades and in standardizing screw pitches and diameters, for example. And the Bureau of Fisheries managed to get enough voluntary compliance with guidelines that a number of depleted fishing grounds began a recovery. Furthermore, the Department's universal focus on waste elimination forwarded awareness of and sometimes even progress in conservation of resources such as coal and timber in spite of the fact that Hoover's major goal was to promote increased consumption through waste elimination. On the other hand, Hoover's emphasis on providing information and advice and allowing voluntary cooperation failed in other very important domains—especially in controlling oil pollution and conserving petroleum supplies.

Hoover's reputation as an unbiased technical expert and his very real ability as an administrator played a major role in one additional area that interested him deeply before he became president in 1929. One of his major interests was in the efficient exploitation of water resources. His view, expressed in a Seattle speech in 1926 was that, "Every drop of water that runs to the sea without yielding its full commercial potential returns to the nation is an economic waste" (Clements 2000, 79). For that reason he urged the development of long-range plans for the development of every major and minor river system to maximize its potential for use, not just in terms of irrigation, power production, and flood control, but in terms of recreational use as well.

Since interstate waterways were under the authority of the Department of Interior rather than the Department of Commerce, Hoover would ordinarily have had little say regarding this resource. But Interior had had virtually no success in producing a plan for developing the Colorado River because of interstate conflicts over water and power allocation, so Hoover became involved. In 1921 Congress authorized the creation of a Colorado River Commission that would include representatives of the states affected presided over by a "neutral" expert who had the ability to understand geological, hydrographic, and construction data as well as the ability to get agreement from the representatives of seven states and the respect to get Congress to go along with the Commission's plans.

Harding decided that only Hoover could do the job. The Commission began its work in January of 1922, and under Hoover's leadership they arrived at a preliminary agreement by November 24 of that year. A variety of exigent circumstances delayed final approval of the plan, which included the construction of a huge high dam at federal expense in the Colorado to produce power and water for irrigation as far away as southern California. Ironically the final bill creating Hoover Dam was signed into law in June of 1929 by then President Herbert Hoover.

His success with the Colorado Commission led Calvin Coolidge to appoint Hoover in 1924 as chair of the Saint Lawrence River Commission which involved international cooperation in creating a waterway to connect oceangoing ships with the Great Lakes and to generate power. Hoover discovered that the Canadian side of Niagara Falls was eroding faster than the U.S. side so that more and more of the water was flowing through the Canadian side, undermining the aesthetic features of both sides and reducing the power generating capacity of the American side. He then got the State Department to bring these issues to the attention of the Canadians, who produced a simple and cheap engineering plan to reduce erosion, preserve the aesthetics, and equalize the flow over both sides of the falls, satisfying all parties.

Hoover's confidence that engineers and engineering knowledge could bring accommodation between opposing sides on almost any issue was brought into question in connection with another project that he was involved in before he became president. In 1920 he was appointed to a "Superpower" committee to try to establish a coordinated power grid for the northeast section of the country. In spite of clear evidence that it would lead to conservation of the fuel and water resources of the country, conflict between public power advocates and private power advocates as well as between those who thought federal oversight was most sensible and those who insisted that the various states should have oversight produced an impasse.

In 1927 a huge and destructive flood in the Mississippi Valley called forth the need for both a massive relief effort and a remediation of the flood control measures that had been put in place by the Army Corps of Engineers over the previous seventy-five years. Once again Hoover was called on to coordinate relief efforts and reconstruction, and once again he was remarkably successful. Some 154 temporary housing camps were established to provide housing for 325,000 persons. These camps provided modern sanitation, running water, electricity, and medical care; but Hoover also organized recreational and educational opportunities for the residents and established a system for residents to maintain a clean and orderly environment (Clements 2000, 116). In addition, he arranged for both federal and private moneys to be allocated for rebuilding; but like other engineers, he paid little attention to how the money was disbursed; so virtually all went to replace major capital assets and none went to poor tenant farmers.

Republicans were especially enthused about Hoover's accomplishments, praising him as "an unprecedented organizer of private citizens for public purposes," and as someone who embodied a "precise fusion of engineering and business ability" (Clements 2000, 125). His public image, already glowing as a result of his work in Europe, as food czar during WWI, and on the Colorado River Commission and St. Lawrence River Commission, became even greater. As a consequence he was chosen to be the Republican candidate for president and was elected to the presidency in 1928.

Hoover's term as president was dominated by concerns arising out of the Wall Street Crash of 1929 and the deepening depression over the next few years. To those concerns Hoover brought his long-term commitment to reliance on expertise, to voluntarism, and to fiscal conservatism. In his early days as Secretary of Commerce Hoover had responded to a minor recession in 1920 by establishing a Committee on Unemployment and Business Cycles, which had recommended the use of public works to counter unemployment during business downturns. As the Great Depression deepened, he did begin to promote the use of public works to absorb some workers, putting in place a number of programs that could be expanded under the New Deal. But Hoover was willing to promote only those public works that would be income producing and which would "not place a strain on the taxpayer and not necessitate government borrowing" (Clements 2000, 175). So his public works projects were largely limited to the construction of Hoover Dam and other Reclamation Projects that were expected to pay for themselves. Once again the upshot was that Hoover's relief policies tended toward capital intensive rather than labor intensive projects such as reforestation, so they did relatively little to reduce unemployment or shift moneys to those most in need (Clements 2000, 182–183).

Hoover's other major effort with respect to major social issues also grew out of his engineering expectation that getting the facts right was critical and possibly even adequate to solving social problems. He established the President's Research Committee on Social Trends composed of experts in economics, sociology, political science, and public health to recommend "correctives to [an] undiscriminating emotional approach and to [an] insecure factual basis in seeking for constructive remedies of great social problems" (Clements 200, 140). In 1933 this group produced a two-volume overview of its work and twelve additional detailed volumes, many of which are still available as reprints in 2013.

Scientific Managers and Public Administrators as Technocrats

Before going on, I would like to quickly review the attitudes associated with Scientific Management and its appropriators for public administration in connection with the six attitudes defining technocrats according to Robert Putnam.

> 1. Above all, the technocrat believes that *technics must replace politics and defines his own role in apolitical terms. . . he has great confidence in the possibility of solving the problems of society by a scientific approach.*

Though such attitudes may have preceded Scientific Management in movements such as Saint Simonian socialism in Europe, they were articulated most clearly and compellingly in twentieth-century America by engineers, many of which were associated with Scientific Management. Even when they admitted that scientific administration should serve political ends, they often did see their own roles as apolitical and they almost uniformly showed confidence in the ability of scientific investigation to solve almost any kind of problem. The major early advocates of public administration argued that major policy directions should be arrived at politically, but they generally insisted that the methods for achieving political goals should be left to administrative experts. Even when elected as public officials, technocrats such as Herbert Hoover often depended heavily on expert apolitical advice. Thus, when the report of his President's Research Committee on Social Trends appeared in 1933, Hoover wrote a foreword in which he said, "It should help all of us to see where social stresses are occurring and where major efforts should be made to deal with them constructively" (President's Research Committee on Social Trends, 1933, i).

> 2. *The technocrat is skeptical and even hostile toward politicians and political institutions.*

By the end of WWI, at least some of those associated with Scientific Management saw political institutions as unable to cope with the modern world, as exemplified by Henry Gantt's views. In addition, they often argued that politicians were corrupted by moneyed interests, although most insisted that administrators should be subservient to public demands.

> 3. *The technocrat is fundamentally unsympathetic to the openness and equality of political democracy.*

Virtually all Taylorites saw technical expertise as best able to guide public policy, and when they were explicit about democracy, they either redefined it to conform to natural laws or assumed that public opinion, sufficiently well informed, would accord with their views. This rarely meant that they were openly unsympathetic to political democracy, but it did often mean that they were frustrated by the presumed equality of individuals and by the failure of publics to listen to the "facts" which they presented. Hoover's tendency to work around congressional restrictions on his authority as "Food Czar" provides a clear illustration of this tendency.

> 4. *The technocrat believes that social and political conflict is, at best misguided, and at worst, contrived.*

This seems to be a corollary of the assumption that social problems can be resolved by appeal to facts rather than opinions, a view promoted by those associated with scientific management. Moreover, it was a central premise of Scientific Management that it should be possible to harmonize the interests of different groups and that the way to do so was through the techniques of Scientific Management. Hoover's roles in the Colorado River and St. Lawrence projects illustrates this view.

> 5. *The technocrat rejects ideological or moralistic criteria, preferring to debate policy in practical, "pragmatic terms."*

Many of the Scientific Management practitioners had studied the American pragmatists, Charles Sanders Piece and George Dewey, agreeing that practical considerations were paramount and always trumped ideology or any non-consequentialist moral claims. This attitude was closely associated with their claim to objectivity and neutrality.

> 6. *The technocrat is strongly committed to technological progress and material productivity; he is less concerned about distributive questions of social justice.* (Putnam 1977, 385–387)

In all of their arguments, advocates of Scientific Management held material productivity up as a social value which was more important than the interests of either wealthy capitalists or the laboring classes. Though they consistently assumed that all elements of the public would be well served by increasing productivity, none showed any specific interest in distributive justice. All of these characteristics were illustrated, for example, in the career of Herbert Hoover.

Thus, to the extent that the vision of production and public service promoted by Scientific Management spread beyond the shop floor between 1911 and the 1930s—it carried almost all of the features of the technocratic mentality as defined by scholars in the 1970s.

Scientific Management and Modernism in the Arts and Literature

Modernism as an artistic and literary movement grew in parallel with the Scientific Management movement, beginning around 1911 and peaking in the 1920s to just before WWII depending on location. It was a response to rapid industrialization and the machine age. Though more a European and Latin American movement than one initiated within the United States, I introduce it here because of its close association with many of the ideals and techniques of Scientific Management—especially those associated with efficiency, standardization, commitment to the general good, the separation of planning from execution, the use of new and improved materials, and emphasis on abstract, scientific foundations for actions rather than traditions of craftsmanship. In addition, within the visual arts in particular—architecture, sculpture, and painting—industrial and state sponsorship began to replace the patronage of the very wealthy. And the new patrons were predisposed toward the values and techniques of Scientific Management.

Architecture was the art most intensively influenced by Scientific Management, and that happened most extensively in Germany, France, and the Soviet Union, even though the Chicago architect Louis Sullivan can in many ways be considered the first of the modernist architects. Long after Sullivan designed and built his first skyscraper around 1884, American architectural education remained in the Beaux-Arts tradition which focused on the use of classical and symmetrical forms or it fell under the influence of Frank Lloyd Wright's opposition to mechanism, focus on integrating buildings into their unique environmental settings, and emphasis on traditional building materials and regional traditions of craftsmanship. Thus, architectural modernism's replacement of symmetry with the repetition of standardized structures independent of the environment, emphasis on the close relationship of form to function, and its embracing of mechanical forms could grow only slowly in the United States.

Though it had variants in different places, Architectural Modernism everywhere had such close relations with Scientific Management and the professionalization of engineering that it is worth emphasizing those commonalities before considering local differences. Like Scientific Management, Modernist Architecture rejected traditional ways of doing things in favor of ways grounded in scientific knowledge. Peter Behrens, the mentor of many of the greatest modernist architects including Mies van der Rohe, Walter Gropius, and Le Corbusier, warned against "imitation of handcraftsmanship" (In Guillén 2006, 27).

Traditional Beaux-Arts architecture had looked to the past—especially to the classical world and the Renaissance—for its forms, and it understood its primary function as "adorn[ing] the edifices raised by man, for whatever uses, that the sight of them may contribute to his mental health, power, and pleasure" (Ruskin 1891, 7). Modernist Architecture, on the other hand, began by considering the function of a structure then argued that form should follow function. Thus, the Italian modernist architect, Giuseppi Terragni wrote that, "a house can in a certain way be compared to a machine and must be constructed so that every one of its parts serves a precise purpose. There should be nothing there that is useless or superfluous, because, like a machine, this will end up hindering its functioning" (In Guillén 2006, 30).

The last sentence in Terragni's statement implicitly emphasizes another concern of Scientific Management—that is, the notion of efficiency. Many modernist architects were very explicit in their emphasis on efficiency. The Mexican architect, Juan O'Gorman, promoted "maximum efficiency and minimum effort" and sought to push that principle down to the level of each space in a design, and Gregori Warchavchick sought to provide "maximum comfort for minimum cost" (Guillén 2006, 22). Turning from houses to office buildings, Mies van der Rohe, perhaps the greatest of German Bauhaus architects, wrote, "the office building is a house of work, of organization, of clarity, of economy. Bright wide work rooms, easy to oversee, undivided except as the organism of the undertaking is divided. [The architect seeks] the maximum effect with the minimum expenditure of means" (In Guillén 2006, 23). In order to achieve efficiency, modernist architects also insisted, as did Scientific Management, that forms must be chosen scientifically, making "full use of modern technology. . . and a scientific approach to human needs and uses" (Guillén 2006, 14). Charles Éduard Jeanneret, better know as Le Corbusier, argued that only the "scientific principles of analysis, organization, and classification" could bring order to architecture (In Guillén 2006, 19), and in 1926 the Soviet architect Nickolai Ladovsky established a psychotechnical laboratory to measure various human characteristics in order to better create spaces for human uses, arguing that while psychophysics, "cannot create artists . . . it can give them all a solid starting point from which they can achieve the aims to which they aspire by the most scientifically correct means" (In Guillén 2006, 26).

In their scientific search for long-term efficiency in architecture many Modernist architects also followed Scientific Management in their emphasis on standardized spaces and forms. Gropius wrote as early as 1910, "To implement the concept of the industrialization of house construction, the company [AEG] will repeat individual components in all of its designs and hence facilitate mass production, promising low costs and easy rentability" (In Guillén 2006, 28). Gropius and May in Germany, Ginzberg in the Soviet Union, Bottani and Griffini in Italy, and the Mexican government all sought to establish a "minimum housing unit" or a "minimum worker's house" (Guillén 2006, 25), and Le Corbusier, sought to establish a standardized "habitation unit" which could be easily reproduced industrially to lower the costs of housing construction in the aftermath of WWI when low cost worker housing was desperately needed. "In order to BUILD," he wrote, "STANDARDIZE to be able to INDUSTRIALIZE AND TAYLORIZE" (In Guillén 2006, 15, capitalization, Le Corbusier's).

Le Corbusier's statement above points to the fact that the parallels between Taylorism and Modernist Architecture were not coincidental, but were often the result of modernist architects' direct and self-conscious appropriation of many elements of Scientific Management. Le Corbusier read *The Principles of Scientific Management* in 1917, citing extensive passages in the paradigmatic manifestos of Modernist Architecture, *Towards a New Architecture* (1923) and *The City of Tomorrow and Its Planning* (1924). Gropius attended meetings of the Taylor Society when he visited the United States. Many Modernist architects used time and motion studies, citing both Taylor and the Gilbreths. They used Gantt charts to schedule projects; and they gave lectures on such topics as "Taylor's Suggestions for the Construction Business" (Guillén 2006, 24).

Finally, although they were as dependant on clients or patrons as the proponents of Scientific Management, the rising advocates of Modernist Architecture were equally insistent that as professionals, their obligations to the general good were more pressing than their obligations to their clients. Distinguishing between what he called the technical architect and the academic—what I have been calling the modernist and the traditionalist—Juan O'Gorman wrote, "The difference between a technical architect and an academic or artistic architect will be made perfectly clear. The technician is useful to the majority and the academic to the minority.. . . . [the first creates] an architecture which serves humanity, [the second] an architecture which serves money" (In Guillén 2006, 108).

In the next chapter we will see that the Scientific Management movement was especially strong in Germany, France, and the Soviet Union. Moreover in Germany and the Soviet Union, between 1900 and 1920, architects either were or came to be trained in engineering-oriented institutions rather than in institutions dominated by the fine arts; so architects were exposed to technical issues associated with machines and new materials as well as the principles of Scientific Management. Furthermore in Germany, industrialists who were promoters of Scientific Management—especially Walter Rathenau and Wichard von Moellendorf at *Allgemeine Elektricitäts-Geellschaft* (AEG)—promoted Modernist Architecture as well, hiring Peter Behrens as the company's chief architect. In the Soviet Union under Lenin's sponsorship of Scientific Management, the state became the primary patron of Modernist Architecture through the Creation of the Moscow Higher State Artistic and Technical Workshops in 1920.

Modernist Sculpture followed a path very much like that of Modernist Architecture, in large part because it was often carried out by the same persons and/or taught in the same institutions. But Modernist sculpture, whether intentionally or by accident, some times revealed a darker side to scientism and technocracy. The central symbol of the Science pavilion at the 1933 Chicago World's Fair, for example, was a sculpture by Louise Lenz Woodruff, which showed a huge Robot pushing a reluctant man and woman into the future. Officially linked to the Pavilion's theme, "Science Advancing Mankind" (Ganz 2000, 783), the statue seemed even more applicable to the secondary theme: "SCIENCE FINDS—INDUSTRY APPLIES—MAN CONFORMS" (Pursell 1995, 230).

Literary Modernism and its close relation, Cinematographic Modernism, were concerned with the same issues of industrialization and mechanization, but they exhibited a much greater range of attitudes toward their implications, shading toward much more overtly critical attitudes. At one end of the spectrum were such friendly and didactic authors as H.G. Wells, who was a great admirer of Henry Ford's system of production because "not an inch of factory space is to be wasted, not a moment of time, not a fragment of scrap, not an ounce of physical strength nor of mental effort" (In Cobley 2009, 169), and Harold Loeb, whose *Life in a Technocracy: What It Would Be Like* offered a paean to Henry Gantt's vision of Scientific Management, with its reduction of the role of traditional politics to a superficial veneer (Loeb [1933] 1996). Loeb's chapter on government, for example, begins with: "Ad-

Figure 2.1. The "Fountain of Science" sculpture at the entrance to the science pavilion at the 1933 Chicago Century of Progress Exhibition. The robot pushing two apparently resisting figures forward into the future brings to mind one of the Hall of Science mottos: "SCIENCE FINDS—INDUSTRY APPLIES—MAN CONFORMS." Source: Century of Progress Records, 1927–1952, COP_17_0009_00288_026, University of Illinois at Chicago Library.

ministration in a technocracy has to do with material factors which are subject to measurement. Therefore, popular voting can be largely dispensed with. It is stupid deciding an issue by vote or opinion when a yardstick can be used" (75), and it concludes with the following paragraph:

> Political government may be retained in a technocracy. Its function would be showmanship. The routine of its executives would be made up of receiving distinguished guests, laying corner stones, making speeches about the rights of man, American initiative, justice. Its offices would be elective, thereby titillating the egos of those who like to think they are running things. Prominent clowns will, doubtless, be frequently elected. Incumbents will not be released from their productive duties. In fact, they will surely make a great to-do about them. How much more dignified would our chief executive have been if he had spent four hours a day steering the scoop of a steam shovel! And how much more useful! (102–103)

Life in a Technocracy stands as a typical but stylistically superior representative of the more than one hundred American technological utopias discussed by Howard Segal in his *Technological Utopianism in American Culture* (Segal 1985).

At the other end of the literary spectrum were such masterpieces as Aldous Huxley's *Brave New World* (1932) and Sinclair Lewis's *Babbitt*, (1922) which railed against the perceived loss of individualism brought on by standardization and rigid planning by technocratic elites, as well as Kurt Vonnegut's *Player Piano* (1952) which explored the potential psychological fallout from technological unemployment and the creation of class warfare between an engineering class and a redundant working class.

With few exceptions, early films reflected negative attitudes toward the modern industrial emphasis on efficiency and technological innovation. Charlie Chaplin's *Modern Times* (1936) exposed the workers' fear of the endless speed-up of the production line and pilloried Scientific Management's emphasis on efficiency, mechanization and standardization in its segment on the "feeding machine." Buster Keaton's *One Week* (1920) offered a critique of the modernist architects' notion of industrialized housing when a prefabricated home kit from the Portable House Company comes with its parts mis-numbered and when moving the house from its assembly site to its permanent location leads to the destruction of the house. Finally, Fritz Lang's *Metropolis* (1927) explored and condemned the new class tensions between technocrats and workers which seemed to be replacing the old class warfare between capitalists and proletarians as well as fears about the replacement of humans by soulless robots.

It is quite clear that the initial motives of most advocates of Scientific Management had been relatively benign, even admirable, in that they sought the mutual benefit of owners and workers alike. When organizations such as the Watertown arsenal or Ford Motors used elements of Scientific Management to exploit workers, Taylor and his colleagues admitted that their techniques could be misapplied for undesirable purposes; but they laid the blame on those who misused the techniques rather than on the techniques themselves. Yet there was a sense in which those techniques took on a life of themselves which belied the initial benign intentions of Taylor and his close colleagues such as Morris Cooke. Even Taylor saw his system of management as somehow more important than the ends to which he initially sought to apply it. "In the past," he wrote, "man has been first. In the future the system must be first" (In Kanigel 1997, 19). And in his treatment of Schmidt, Taylor certainly seemed to authorize the inhumane treatment of a human being in the name of increased productivity.

If these attitudes had remained applied only to the factory, they would have deserved condemnation for their treatment of workers; but they became ubiquitous in virtually all domains of life between 1911 and 1930. According to Evelyn Cobley, author of *Modernism and the Culture of Efficiency,*

> scientific management adapts the way a virus does, fitting in almost everywhere. . . . The impact, not always recognized, of this 'virus' on both individuals and the social order was enormous. Once scientific management spread from the factory to public and private offices, the most significant consequences were often the disavowed reification of the individual human consciousness and the mostly overlooked depoliticization of the public sphere. (Cobley 2009, 113)

Unlike such mid-Victorian authors as William Morris, whose *News from Nowhere* (1887) harkened back longingly to a idyllic mediaeval and pre-mechanical culture, or Samuel Butler, whose *Erewhon* (1872) looked forward to a culture that had completely rejected the machine, most modernist novelists were unwilling to reject the obvious material benefits of industrialization, mechanization, and technological progress. They were, however, profoundly disturbed by the tendency to replace traditional values with an obsession with productivity and efficiency as ends in themselves as well as by the total dependency on mechanization which industrialization seemed to demand.

Though a few Modernist writers, such as D.H. Lawrence, saw the irrational elements associated with eroticism as offering a powerful antidote to the supposed rationality associated with the new ideology, many modernist novels including John Dos Passos's *The Big Money* (1936) and Sinclair Lewis's *Babbitt* (1922) condemned the internalization of the values they associated with Taylorism and "Fordism," without offering any escape. Dos Passos sticks close to the factory context in which Taylorism produces Ford's production line with its deadening monotony: "reachunder, adjustwasher, screwdown bolt, shove in cofferpin, reachunder, adjustwasher, screwdown bolt, reachunderadjustscrewdownreachunderadjust, until every ounce of life was sucked off into production and at night the workmen went home gray shaking husks" (In Banta 1993, 4–5). He condemns "the college students with stopwatches and diagrams, tabulating, [and] standardizing," and argues that the workers viewed the stopwatch as symbolizing "a hideous invasion of privacy, an oppressive all-seeing eye that peered into their work lives, ripping at their dignity." He quite correctly accuses Taylor of devaluing "the great mass of traditional knowledge which in the past has been in the heads of the workmen and in the physical skill and knack of the workmen "(In Cobley 2009, 50–51). But above all, he hates the way in which Taylor managed to get workmen such as Scmidt to internalize the ideology of efficiency and to see themselves as beneficiaries rather than as victims of the worship of efficiency.

Sinclair Lewis was more concerned with how citizens beyond the factory internalized Taylor's obsession with productivity and efficiency. In *Babbitt*, for example, Babbitt argues that correspondence schools are superior to universities because they are more efficient and don't waste valuable time on "studying poetry and French and subjects that never brought in anybody a cent" (Lewis 1922, 65). Subsequently, Babbit condemns the universities, whose irresponsible teachers and professors [will not] help us "by selling efficiency and whooping it up for rational prosperity" (Lewis 1922, 143). Alas, pleas for greater emphasis on STEM education in 2015 sound all too much like those of Babbit's son in 1922.

Just as Taylor had seemingly lost sight of the fact that efficiency was a means to the independent end of well-being for both workers and owners, Babbitt had lost sight of the fact that selling was just a means to provide for the needs of a customer. He had become dedicated to "the cosmic purpose of Selling—not selling anything in particular to or for anybody in particular, but pure Selling." Similarly, he had lost sight of any reason for "hustling" and had become committed to the practice of "hustling for hustling's sake." After lunch he "hustled back to his office . . . with nothing much to do except to see that the staff looked as though they were hustling" (Lewis 1922, 117). Finally, Babbitt gave up any sense of individuality to embrace standardization in all of its possible dimensions. His clothing consisted of a completely undistinguished "standard" gray suit and "standard boots," his bedroom reflected the "standard designs of the decorator," which included the "standardized electric bedside lamp" to aid him in reading the "standard bedside book" (Lewis 1922, 11). Even the mid-sized city in which he worked was indistinguishable from any other city in Oregon or Georgia, Ohio or Maine, Oklahoma or Manitoba. Nowhere in *Babbitt* is there the slightest hint that he could imagine escape from what Lewis clearly saw as the stultifying conformity promoted by Taylor's system.

Huxley's dystopian *Brave New World* and his successor utopian *Island* contain a more subtle, and to my mind a more important analysis of the dangers and opportunities associated with such movements as Scientific Management. His criticisms are not of the application of science to human activities but of the application of a science that is too narrowly focused on a very small number of relevant variables, which is applied in a routinized and unthinking way, and which is obsessed with material well-being, rather than with "truth and beauty."

In both *Brave New World* and *Island*, for example, inhabitants are directed in their lives by science, but in the first, according to world controller Mustapha Mond, "science is just a cookery book, with an orthodox theory of cooking that nobody's allowed to question, and a list of recipes that mustn't be added to except by permission of the head cook" (Huxley 1946, 270). In much the same way, Taylor and his disciples established the techniques used in Scientific Management, stipulating exactly how they were to be used and condemning variations suggested by Harrison Emerson and other efficiency experts.

In *Island* on the contrary, citizens are taught to be attentive to the application of all knowledge and to be wary of "the systematic taking of unanalyzed words much too seriously" (Huxley 1962, 37). Furthermore in *Island,* children are taught from the beginning of their education about the connectedness and complex interdependence of all things: "Never give children the chance of imagining that anything exists in isolation" (Huxley 1962, 217). All things must be analyzed within their broad context, so there can be no universal method for scientizing material production, education, managing public policy, and so forth.

Starting from the techniques that could make material production more efficient and which thus focused on consumer demands distorted the basic goals of science. Thus wrote Huxley, "Ford [read also Taylor] himself did a great deal to shift the emphasis from truth and beauty to comfort and happiness. Mass production demanded the shift. Universal happiness keeps the wheels steadily turning; truth and beauty can't" (Huxley 1946, 273).

Chapter Three

Taylorism Goes Global

The Spread of Technocratic/Scientistic Ideologies, 1910–1943

Taylor's reputation in Western Europe was built initially on the spectacular success of his development of a hard steel alloy that retained its hardness and edge at very high temperatures and thus allowed for increases in metal cutting speeds of up to 300 percent. Displayed at the Paris Exposition of 1900, Bethlehem Steel's monster lathe using Taylor-White alloy tools was the hit of the show and led European industrialists to favorably consider many of Taylor's other efficiency-increasing techniques. Except for a few early German converts, those who introduced Taylorite ideas outside of the United States, tended to be flexible and to be willing to appropriate elements of his system without insisting that the system be adopted all-or-nothing. Moreover "Taylorism" was frequently modified to fit different cultural circumstances. And Taylorite ideas penetrated different cultures to different extents. In some it remained a feature of industrial engineering culture, while in others it spread into political ideology and the management of public as well as private entities as it had in Progressive Era America.

France

France offered a particularly welcoming environment for Scientific Management for a number of reasons. First, engineers trained at the *École polytechnique, École central, École des mines, École des ponts et chaussees,* and the *Conservetoire des arts and metiers,* already played major leadership roles in both industry and public life, so technical knowledge already commanded widespread respect. Second, most of those trained in the *Grande Écoles* by the beginning of the twentieth century had extended exposure both to Auguste Comte's Positivism, which urged the application of scientific knowledge to all activities, including social ones, and to Saint Simon's emphases on the centrality of productivity in industrial society as well as the efficiencies to be gained by centralized planning.

Henri Le Chatlier, famous chemist and professor at the Sorbonne, had visited the high speed steel exhibit and was much impressed. When he established the *Revue de Métallurgie* a few years later, he praised Taylor's work. Taylor responded by sending him a metal cutting paper, which Le Chatlier asked permission to publish. When Taylor agreed, he also sent a copy of *Shop Management,* which had a huge effect on the Frenchman, who wrote, "I had not foreseen the possibility of extending the domain of science over all the realm of industry, including questions of organization, commercial questions, labor questions, etc . . . From that day on I felt myself . . . an apostle of the Taylor system" (Kanigel 1997, 403). Moreover as a number of Taylor's technical papers were translated into French through 1907, Le Chatlier published extensive sections in *Revue de Métallurgie.*

A second early convert to Taylorism, Charles de Frémenville, was a friend of Le Chatlier, the ex chief-engineer of the Paris-Orléans railroad, and a engineer in the early French automobile industry. Introduced to Taylorism through the *Revue de métallugie*, Frémenville only became committed to promoting Scientific Management after a meeting with Taylor in 1912, shortly after the Eastern Rate Case brought him to prominence. Frémenville returned to France to begin a series of lectures on Taylorism to industrial societies and to write articles on Scientific Management for a variety of French journals (Merkle 1980, 150).

After reading the French version of *Shop Management* in Le Chatlier's *Revue de métalurgie*, Georges Ram, mechanical director at the Renault plant in Billancourt used time and motion studies to set piece rates for a trial group of 130 workers. Though many workers quit, Ram reported a 100 percent increase in productivity in his own article for the *Revue* in 1909. Soon after, Louis Renault was on his way to the United States to meet with Taylor. Like many of Taylor's American clients, Renault insisted upon pushing forward with time-study based wage setting without going through the lengthy preparatory steps that Taylor recommended, leading to two costly and bitter strikes which delayed the implementation of Scientific Management until WWI. Then the need for the rapid production of military vehicles with an increasingly unskilled work force pushed forward a less abrasive implementation of many scientific management techniques for increasing efficiency.

The experience of the Michelin Tire Company with Scientific Management was much more favorable than that of Renault, in large part because Marcel Michelin initially appropriated less alienating elements of Scientific

Management. Initial successes led to a long-standing effort by the Michelin family to promote Scientific Management. Again it was Le Chatelier who convinced Andre and Eduard Michelin, principle owners of the Michelin Company, to read Taylor's *Shop Management* and *Scientific Management.* They saw the potential applicability of Taylorite techniques to their business and after meeting with Taylor during his 1913 trip to Paris, they sent Andre's son, Marcel, who knew English and was head of testing at their Clermont Ferrand plant, to America to learn more. Taylor turned the young man over to H. King Hathaway, a collaborator, because Taylor's wife was sick, and Hathaway took him to see the Tabor Manufacturing Company, Link-belt, Towne and Yale, as well as a few other sites where Scientific Management was being successfully implemented. In addition they visited a small Michelin Tire factory in nearby Milltown, Pennsylvania.

Michelin was interested in how Taylorite ideas could be used to improve productivity of the small (25–30 worker) machine shop at Milltown and then be scaled up to the 3,500 person machine shops at Clermont Ferrand. Hathaway acknowledged that the Milltown plant was already well managed and that improved productivity would depend on very detailed analyses of every operation. He also sent Marcel home with a detailed twelve-page outline, edited by Taylor, regarding how to deal first with the Milltown plant and then with the larger Clermont Ferrand plant. The young man was impressed and wrote to Taylor, saying, "Now I understand a great deal better what is meant by 'Scientific Management' and I realize what an enormous advantage can be obtained by working in accord with such a method. I am going back to France now, and I am quite prepared to discuss, with my uncle, the way we will apply your method to our particular branch of industry" (Tesi 2009, 12).

The Michelins decided to initiate elements of Scientific Management immediately in France, but they drew very selectively from Hathaway/Taylor's plans and incorporated elements that were not strictly derived from Taylor. For example, they engaged in what Taylor derided as "welfare activities," including the provision of low cost housing to employees. The coming of the First World War accelerated their application of Taylorite methods for organizing work flow in their plants for making airplanes and gas masks. Immediately after the war the Michelins reorganized their machine shops using redesigned special tools, improved systems for delivering tools to where they were needed, Taylorite planning methods and accounting procedures, increasing output while reducing costs. In addition they created an employee suggestion system which was consistent with Taylor's ideas but whose details had been suggested by Hugo Munsterberg in his 1913 *Psychology and Industrial Efficiency*. In 1919 the Michelins went well beyond Taylor's views when they began building housing for workers using their own workforce organized along Taylorite lines. Working hours per house were driven down from 11,643 in 1920 to 5,548 in 1925, allowing Michelin rents to be much lower than market levels.

Because of their success in applying Scientific Management techniques, the Michelin brothers and their most important managers established a foundation which funded the creation and operation of *The Comité Michelin* to promote Taylorite ideas. With the support of a number of leading engineers, the *Comité* established a special collection in the Library of *the Société d'Encouragement pour l'Industrie Nationale*; it paid the salary of a special librarian to constantly update materials on Scientific Management; and it funded a series of seminars and internships to train young engineers in Scientific Management issues. The death of Andre Michelin in 1931 brought an end to *The Comité Michelin*, but in the meantime the firm began to finance its own training school in Scientific Management for young engineers, picking off the cream of the crop for its own workshops. In addition, between 1928 and 1933 it published a popular magazine titled *Prospérité,* promoting Scientific Management (Tesi 2007).

Early wartime successes in using the techniques of scientific management, not just by Michelin, but by many companies, led Georges Clemanceau, radical republican prime minister of France and minister of war from 1917 to 1920, to request that all war plants consider Taylorite techniques and establish Taylorite planning departments in 1918 (Maier 1970, 37–38). Soon after, he ordered that Taylorism be introduced into government departments as well" (Kanigel 1997, 493).

French Scientific Management, though it often referred to Taylor and his work, was developed largely by Henri Fayol, who admired Taylor but changed some of his emphases. Fayol was even more sympathetic to workers and their technical skills than Taylor and argued strongly that workers were rewarded by things other than money, including a sense of participation in shaping their jobs. Perhaps more importantly, unlike Taylor, who started by analyzing the physical characteristics of worker activities and worked his way up to considering planning and management, Fayol worked from the top down, emphasizing management functions and working down to shop floor phenomena. This approach was clearly more appealing to French Industrialists who came from a tradition of hierarchically structured industry and society.

Regardless of their differences, both Taylor and Fayol argued that increased productivity would come from the application of expert scientific knowledge to virtually any task. Both fostered an ideology of administrative rather than political solutions to problems and argued that tensions between workers and owners could be resolved by the application of disinterested expertise. That is, they were equally technocratic.

Germany Before the Rise of Hitler

Germany also provided a fertile ground for Taylorism and Scientific Management at the beginning of the twentieth century. First, though Germany was slow to industrialize in one sense—most manufacturing continued to be done in the old personalized master-apprentice context in which firms typically had five or fewer employees well into the 1880s—by 1900 Germany was probably the most scientifically and technically advanced country in Europe. *Technische Hochschulen* began to spring up in 1860, and engineers had already formed the *Verein Deutscher Ingenieure* by 1859. Second, from the time of the Cameralists in the early eighteenth century, planning based on scientific analyses had been important within government bureaucracies; so a dependence on technical knowledge was nothing new. On the other hand, the German army offered a model of a highly disciplined and efficient organization which businesses and industries could emulate; so German responses to Taylorism largely ignored his emphasis on functional management in favor of top-down, military type management.

When production moved into large-scale units, the older personal styles of management most common in Germany were inadequate. As increasing industrial competition came especially from American firms, there was a growing fear of an "American peril" and a tendency to look to America for techniques to improve efficiency. The first generation of German Taylorite promoters were engineer members of the *Verein Deutscher Ingenieure* who became interested in Taylor as a result of his high-speed steel experiments. Georg Schlesinger, leader of the *Verein Detcher Ingenieure*, for example, argued that the high speed steel demonstrations were "a landmark in the history of mankind" and he undertook a leading role in spreading Taylorism in Germany where *Shop Management* was translated in 1904, just one year after its appearance in English (Kanigel 1997, 405). Many German engineers traveled to the United States to study American production practices, and a surprising number of these were recruited into management by major firms. Initially these engineers attempted to install the complete Taylor system emphasizing scientifically established piece-rate compensation with the same labor opposition and limited success that appeared in the United States, but many retained their positions and refined their approaches to suit German conditions.

Unlike the case elsewhere, in Germany, Taylor's major writings were translated by academic engineers, who promoted Taylorite ideas among intellectuals as well as among practicing industrial engineers, and they were picked up by a brilliant popularizer who was well connected in aristocratic and government circles. Thus Scientific Management in Germany had multiple important audiences, not only among industrial engineers and government bureaucrats; but among artists and architects as well.

Professor Adolph Wallichs translated *Shop Management* in 1904 after he had visited Bethlehem Steel and become a strong advocate for the application of Taylor's methods in Germany. And Dr. Rudolph Roessler translated *The Principles of Scientific Management* in 1914. But the most important early promoter of Taylorism was Wichard von Moellendorf, who obtained a degree in mechanical engineering from the Technical University of Charlottenberg in Berlin in 1906. A friend, Walter Rathenau, hired him to organize a metals laboratory at a cable factory owned by the Rathenau family-run *Allgemeine Elektrizitäts Gesellschaft* (AEG) in 1908, and the two continued to work together both at AEG and later, in organizing materials supplies for the German war economy during WWI. Around 1914, von Moellendorf began publishing articles on Taylorism in major intellectual journals, such as *Die Zukunft (The Future)* and *Die neue Rundshau (The New Review),* focusing on Taylor's critique of the "egoistic profit seeking" of owners and arguing that Taylorism would not only serve the public interest by harmonizing owner and worker interests, it would achieve efficiencies that would increase productivity and lower consumer prices.

In 1916 von Moellendorf published *Deutsche Gemeinwirtschaft (The German Communal Economy)* in which he proposed an economy directed by society in which engineers would play a central role as priests of "efficiency [and] applied objective causality" (Guillén 1990, 11). Though von Moellendorf was a socialist, his idea of the communal economy and the emphasis on engineers as leaders was appropriated by neo-conservatives and made the foundation of a series on the German Communal Economy that sold over 44,000 copies between 1917 and 1920 (Guillén 1990, 11).

Ultimately, von Moellendorf's reading of Taylor became the foundation for a powerful German fusion of technology and Idealism that fit beautifully with Nazi ideology. In Heinrich Hardensett's *Der kapitalische und der technische Mensch* of 1932, the basic line of argument was laid out. Capitalists, at home in a world of abstractions and focused on profit so obsessively that it is "monomaniacal to the point of pathology," introduced Taylorism into German industry solely to increase private profits, thus eliminating all comradely and joyful elements from work. "The Technical man, on the other hand, is devoted to 'service value' and the production of things, concrete objects which are 'cosmic, eternal, and godly." This involves an essentially creative act of the will, or soul, which was Germanic at its core (Herf 1984, 638–39).

Von Moellendorf became undersecretary in the Ministry of Economics in the Weimar government in 1918, but his ambitious plans to initiate the communal economy, which he admitted could be characterized as "capitalist techniques under the control of society" (Merkle 1980, 190) were rejected in May of 1919 and he was forced to resign from public office and return to metals research. Moellendorf's plan contained something to offend everyone, but part of the opposition undoubtedly came from Taylor's reputation for hostility to and disrespect for labor—reinforced by the Schmidt episode in *The Principles of Scientific Management,* which identified a German worker as "dumm und stark wie eine Stier" (Merkle 1980, 194). Not only was German industrial psychology far enough advanced to challenge the humanity and efficacy of some Taylorite techniques, but German socialism retained an element from Fourier, Saint Simon, and Marx that, in Rathenau's words, insisted that there should be "joy in work" (Merkle 1980, 188), a concept completely absent from Taylor's version of Scientific Management but present in Cooke's.

Ironically, starting late in 1919, German union leaders, led by Gustav Bauer, who was a union member, a member of the Social Democratic Party, and Minister for Labor, began to adopt Taylorism. Bauer explained, following an argument initiated a few years earlier by V.I. Lenin, that under capitalism,

> Taylorism stood in opposition to the worker. Labor feared that the capitalist and not itself, would be the beneficiary of the new method of work. Now that the democratization of Germany has ensured an ample economic influence of labor, these objections are not only futile but they also obscure the fact of the possibilities offered by rationalization. (Guillén 1990, 18)

Carl Köttgen, an electrical engineer at Siemans who was converted to Scientific Management while visiting the United States, in turn, converted Carl Friedrich von Siemens to Scientific Management. Together they convinced other important industrialists and the German government to establish a National Board for Efficiency (*Reichskuratorium für Wirtschaflicheit,* RKW) funded and staffed initially by the state, but increasingly by the great cartels, to initiate centralized industrial planning, standardization, and rationalization efforts. The great success of this organization in increasing productivity and lowering costs of production while improving the standard of living for workers produced what has been often called the first "economic miracle" of German industry up until the great depression of 1929 (Merkle 1980, 198). Moreover, it created a huge appetite for Scientific Management or "Taylorismus" literature in inter-war Germany that was fed by over a thousand books and articles (Nelson 1992, 23).

Technocracy and the Third Reich

In 1932 Adolph Hitler ran for the German Chancellorship on a platform that was explicitly anti-efficiency, arguing that the whole German rationalization movement had brought on the depression of 1929 by creating unemployment as a consequence of overproduction. But behind the scenes the economic system of the Third Reich was, as Judith Merkle has argued, "a direct descendent of the rationalization movement and the 'New Economy' proposed by Rathenau and Moellendorf. For that reason it showed strong Scientific Management influence" (Merkle 1980, 219). Equally important, German engineers, some only very distantly influenced by Taylorism, played a significant role in formulating Nazi ideology. Gottfried Feder, for example, was a key figure in developing Nazi propaganda in the period before 1933. He organized a series of pamphlets, including *Nationalsozialismus und Technik: Die Geistigkeit der nationalsozialistischen Bewegung (National Socialism and Technology: the Spirituality of the National Socialist Movement*), which appeared in 1930. This work argued that "the primal life instincts pulsing in technics would help [the Nazis] break free from the fetters of Jewish materialism (Herf 1984, 640)." The Nazi organization for college educated engineers, the *NS Bund Deutscher Technik*, numbered over 50,000, and it published a monthly journal, *Deutsche Technik*, whose editors included Feder as well as Fritz Todt, and later, Albert Speer, who we will hear more about soon. Even the long-standing *Verein Deutscher Ingenieure* produced a monthly journal, *Technik und Kultur,* which from 1923–35, "presented an image of the engineer–artist struggling to beautify the external world in the face of commercial interests bereft of an appreciation or understanding of the . . . foundations of technology" (Herf 1984, 637). The significant aesthetic dimension of this movement certainly transcended Scientific Management and it had a major impact on European artistic culture.

In addition, Hitler was a technophile who was inclined to turn over huge national projects to individual scientists and engineers, who in turn mobilized union members and the great cartels to get work done using only slightly modified Taylorite methods. Robert Ley, a PhD in chemistry, became Reich Organizational Director in 1933. Tasked with what to do about the massive labor movements of the Weimar Republic. Ley outlawed prior free labor unions and established the Labor Front (DAF) to represent all laborers. Then he used a series of moves recommended by industrial psychologists to reduce opposition to his autocratic rule to almost nil. For example, his "Strength through Joy" program provided workers with subsidized vacations, including cruises on two luxury cruise ships, as well as improved housing, and his "Beauty of work" program sought to make workplaces more appealing to workers (Taylor 2010, 14–15). During WWII, less pleasant elements of Scientific Management, including the micro-division of labor made it possible to use foreign forced labor in complex technological tasks.

Fritz Todt, the father of the autobahn system, got his first engineering degree from a *Technische Hochscule* in Munich. After serving in the air force during WWI, he joined the Munich civil engineering firm of Sager and Wörner, where he worked his way up from laborer to technical leader and manager. Todt specialized in road surfacing and earned a doctorate of engineering, writing a dissertation on *The Causes of Defects in Asphalt on Roads* in 1931. In 1932 he became head of the National Socialist Technical Union. In the following year Hitler turned a vague general plan for improving existing roads and constructing new ones over to Todt, who Hitler described as "a man who was suited to transform a theoretical intention into a practical reality" (Tayor 2010, 17). Over the next ten years Todt not only nearly completed the impressive autobahn system, but, according to Albert Speer, "the modest, unassuming technologist gathered into his hands responsibility for the entire German construction industry [including] . . . all navigational waterways and power plants" as well as such special projects as the construction of the Siegfried line, the Olympic Stadium in Berlin and the concentration camps in Germany and Nazi occupied Europe (Taylor 2010, 16–17).

By all accounts Todt was little interested in party politics, but he did combine his technocratic views with elements of German romantic nationalism. Thus when he published a 1937 book, *Germanies Highways* he wrote: "A road is a work of art. It arises from the engineer's creativity. The service of the road purely as a means of transportation is not the only purpose of German road construction. The road must be an expression of the environment, and an expression of German essence" (Taylor 2010, 47). In 1938 the Organization Todt (OT) was set up outside both the military and the Nazi Party to carry out military construction projects. Free of both military and party bureaucracies and answering only to Todt, the OT was incredibly efficient. Todt became Reichsminister for Armaments in 1940, but the OT continued to carry out vital wartime construction projects. According to British intelligence reports, over the five years following 1938, over 1,400,000 men, many of them POWs, worked for the OT, carrying out "the most impressive construction program in modern history" (Taylor 2010, 73).

As Reichsminister for armaments Todt exhibited the technocrat's pragmatic emphasis by imposing a compromise between the generals' requests and industrial capacity, forcing the army to stop requesting endless numbers of short-run specialty items and forcing the armaments industry to focus on coordinating production and supply of key materials (Taylor 2010, 94). As long as an industry completed its assigned tasks, Todt deflected interference from the generals, the party bosses, and government bureaucrats. When the United States entered the war, Todt told Hitler that Germany could not win, and on February 7 of 1942, after an acrimonious six-hour meeting with Hitler, Todt left, depressed. A few hours later the plane he was flying in, probably as co-pilot, crashed with no survivors.

Todt's successor as Minister of Armaments was the much more famous Albert Speer, who was convicted of war crimes at Nuremberg. Following twenty years in prison, he became author of the best selling *Inside the Third Reich.* Like Todt, Speer began his technical education at a *Technische Hochschule* and completed it at a technical university, this time at Munich and then Berlin. In spite of his preference for mathematics, Speer's architect father insisted that he get his degree in architecture. While teaching at Berlin in 1930, Speer heard Hitler speak, and soon after, he joined the Nazi party. Even before the Nazis assumed power, Speer arranged for the renovation of the party headquarters, and when Hitler came to power Speer renovated the Old German Reich Chancellery in Berlin, establishing a close friendship with Hitler. After undertaking a number of other architectural duties, Speer was named General Building Inspector for Berlin in 1937 and he designed and built a series of Führer headquarters throughout German occupied territories while he headed the Beauty of Work program.

Hitler was impressed with Speer's organizational abilities as were many others. Ferdinand Porsche, creator of the Volkswagen Beetle, spoke of his brilliance, and one observer wrote, "Speer was a mediocre architect, but an extraordinary technocrat and organizer" (Taylor 2010, 131). So when Todt died in 1942, Hitler immediately named Speer Minister of Armaments, a post through which he effectively and competently controlled the Reich's economy, admitting his debt to Rathenau's rationalization movement (Guillén 1990, 30).

Though recent authors have argued that Speer was a Nazi through and through, he gave an impression both within the Reich—he chose not to attend Party meetings and did not openly engage in political infighting—and beyond that he was less a Nazi than a technocrat. In a famous 1944 editorial in the London *Observer* the author used Speer as a case study to express both a resignation toward and a growing fear about technocrats and technocracy in general:

> Speer is—in a sense—more important for Germany today than Hitler, Himmler, Göring, Goebbels or the generals. They all have, in a way, become the mere auxiliaries of the man who directs the giant power machine—charged with drawing from it the maximum effort under maximum strain. . . In him is the very epitome of the "managerial revolution.". . . . Much less than any of the other German leaders does he stand for anything particularly German or particularly Nazi. He rather symbolizes a type which is becoming increasingly important in all the belligerent countries: the pure technician, the classless bright young man, without background, with no other original aim than to make his way in the world and no other means than his technical and managerial ability. It is the lack of psychological and spiritual ballast—and the ease with which he handles the technical and organizational machinery of our age—which makes this slight type go far nowadays . . . This is their age; the Hitlers and the Himmlers we may get rid of, but the Speers—whatever happens to this particular special man—will long be with us. (Speer 1970, 442–443)

Keep this editorial in mind, especially as we consider in chapter four what happened during the Cold War in both the USSR and the USA.

The Soviet Union before WWII

It might seem odd that a set of practices designed and implemented to improve capitalist production should find an immediate welcome in the first fully socialist state and one that was as industrially backward as any in Europe; but we have already seen that many German advocates of scientific management emphasized the idea expressed by Taylor, Gantt, and others that they aimed primarily at increasing productivity rather than profits. Capitalists might have been the first to utilize Scientific Management to exploit workers effectively; but its techniques were easily adapted to increase productivity in non-capitalist circumstances.

When he had first learned of Taylorism in connection with the huge strike of workers connected with Renault's 1913 attempts to initiate Taylorite practices at its Billancourt plant, V. I. Lenin had condemned it as "a scientific system to crush the worker" (Peci 2009, 293). But when he was preparing to finish *Imperialism: The Highest Stage of Capitalism* in 1916, his views changed. As part of that project he studied the latest in capitalist methods of worker exploitation, and he was especially impressed by the methods of Scientific Management, which not only managed to increase productivity, but which managed to do so in a way that broke down complex physical and record keeping functions into such simple procedures that any ordinary worker could do them. One of Lenin's notebooks from this time, *Tetrad' beta*, contains long extracts from Taylor's writings with notes expressing his admiration for the many experiments that went into producing high-speed steel and for the ruthlessness of Taylor's effective methods for lowering the costs of industrial production (Merkle 1980, 106). Lenin also studied and admired Frank Gilbreth's use of motion studies—especially because they came from a worker's perspective. Though he did not mention Taylorism by name in his pre-revolutionary *The State and Revolution* which appeared in early 1917, Lenin did argue that the ultimate withering of the state would be made possible by reducing the methods of accounting and control to such simple processes that they could be carried out by ordinary housewives—a fairly clear reference to the claims of Scientific Management (Merkle 1980, 109).

When the Bolsheviks finally came to power at the end of 1917 and Lenin became the de-facto head of the state, the Bolsheviks faced what seemed to be a virtually impossible situation. They were engaged with the Germans in World War I. At the same time they were engaged in a civil war with both the White Russian Army on the Right and with more radical Communist groups on the Left. Critical resources such as the railroads were unuseable for lack of skilled maintenance workers and replacement parts. Industrial capacity—even for the production of essential munitions—was virtually destroyed; and agricultural production had been dramatically decreased for lack of workers.

By March 3, 1918, the Bolsheviks had extricated themselves from war with Germany, and Lenin turned to the issues of industrial and agricultural recovery. On April 1, 1918, he addressed the Supreme Council of the National Economy regarding a decree on labor discipline. In the decree, he argued "we must definitely speak of the introduction of the Taylor System, in other words, of using all scientific methods of labor which this system advances. Without this, it will be impossible to increase productivity, and without that we will not [be able to] usher in socialism" (Bailes 1977, 376 n10). Within a few days Lenin also published his first major article as head of the Soviet government in *Pravda*, the party newspaper. "We, the Bolshevik party have *convinced* Russia, we have *won* Russia from the rich for the poor, from the exploiters for the toilers. It is now up to us to *manage* Russia," he began. In this task of management the first step was to raise the productivity of labor, and to accomplish that the first step was to improve "the emotional and cultural state of the masses," which would be accomplished principally through study groups of the communist party. But there was a crucial second step:

> Higher productivity of labor depends, in the second place, on higher discipline of the toilers, on higher skill, efficiency and intensity of labor, and on its better organization. . . We should immediately introduce piece work and try it out in practice. We should try out every scientific and progressive suggestion of the Taylor System. . . . The last word of capitalism in this respect—the Taylor System—as well as all the progressive measures of capitalism, combined the refined cruelty of bourgeois exploitation and a number of most valuable scientific attainments in the analysis of mechanical motions during work, in eliminating superfluous and useless motions in determining the most correct methods of work, the best systems of accounting and control, etc. The Soviet Republic must adopt valuable and scientific technical advance in this field. The possibility of socialism will be determined by our success in combining the Soviet rule and the Soviet organization of management with the latest progressive measures of capitalism. We must introduce in Russia the study and teaching of the new Taylor System and its systematic trial and adaptation. (In Del Mar and Collons, 1976, 376–377)

In the same month Taylor's piece-rate system for setting wages was adopted to promote efficient production and Lenin insisted that workers who blatantly fought against the practice should be imprisoned (Beissinger 1988, 29). Piece-rate systems slowly worked their way into Soviet industries as techniques for rate-setting were adopted. Although they had to be modified in 1926 to allow for increased wages in critical industries, as piece-rates were abandoned, other forms of individualized incentives were created by the Central Labor Institute—including recognition as a "Labor Hero," or inclusion of one's productive accomplishment in the factory newspaper (Peci 2009, 298).

Early Soviet uses of Scientific Management were not limited to issues of productivity at the factory or even the industry level. Walter Polakov, who had been a member of Gantt's "New Machine," served as consulting engineer to the Supreme Council of the National Economy as it sought during the mid-1920s to produce the first Five Year Plan that would organize the entire economy, and the plan was drawn up on Gantt Charts, the explanations of which Polakov translated into Russian. Thus Scientific Management took on a critical place in National Planning and there was a tendency for some party members to claim that Scientific Marxism and Scientific Management were identical (Merkle 1980, 122–123). Even many of those who opposed the relatively narrow implementation of Taylorite ideas to shop floor activity saw the broader application of Scientific Management to society as a whole as consistent with Marxism. Thus Platon Kerzhentsev, one of the most adamant opponents of the strict application of Scientific Management to the "scientific organization of labor (NOT)" at the shop floor level, argued that, "'To a Marxist'. . . principles derived from the rationalization of production would eventually be applied to every organizational activity, be it schools, the state apparatus, or the Red Army" (Sochor 1981, 252).

In all of his attempts to implement scientific management in Russia, Lenin was supported by the Central Labor Institute (Sochor 1981, 249–259), the creation of Alexei Gastev, a popular "worker poet" and metal worker who brought to the task the kind of romantic enthusiasm for engineering and industry that informed the German *Techniks and Kultur* literature. In his poetry, "men and machines merged: machines were seen as extensions of the human body while people took on the speed and efficiency of their creations, acquiring 'nerves of steel' and 'muscles like iron rails'" (Bailes 1977, 374). Gastev's goal for the Central Labor Institute, formed under the sponsorship of the All-Russian Council of Trade Unions in 1920—"to implant into the character of the peoples of Eastern Europe and Siberia, the searching, daring, persisting American energy and productivity"—certainly had romantic elements. Its structure, however, was grounded in the "functional management" practices promoted by Taylor, who Gastev had encountered before 1914, supplemented by industrial psychology as it had developed in Germany and France. The institute had six divisions, focused on a) cinematographic studies of work processes, b) the development of new machine design, c) the mechanics of work movements, d) energy utilization and conservation, e) industrial psychology, and f) the teaching of scientific management principles (Del Mar & Collons 1976, 384).

Gastev promoted engineering expertise in the perpetually underfunded Central Labor Institute, following the Taylorite notion that the vast majority of industrial jobs could be routinized while a small number of engineering experts established the routines. Writing of the proletarian social psychology incorporated in such a system Gastev argued that, "The psychology of the proletariat is already being transformed into a new social psychology where one human complex works under the control of another," with the workers being controlled by the engineers (Bailes 1977, 378). While this view may have been consistent with Lenin's notion that at least for the time being, while Russia would be governed *for* the proletariat, it would have to be governed *by* an elite vanguard, it was opposed by such leaders of the Proletarian Culture movement as Alexander Bogdanov who insisted that socialism meant the existence of a single class not subject to control by anyone but itself. According to Bogdanov, "Where work demands the direct supervision of an individual person, there will emerge, instead of authority and force, a comradely recognition of competence; and he who in one endeavor was the instructor may then in another follow the directions of a comrade whom he had just been supervising: the organizer and executor change places frequently" (Bailes 1977, 380). Bogdanov and those associated with him hung on to the old Socialist notion that work should involve the satisfaction of human needs for self-expression, while Gastev promoted a culture of work that involved "a postponement of immediate satisfaction which may be called conditioning for work" (Bailes 1977, 384).

Given the demands of modernization and industrialization which had to put productivity before worker satisfaction, Lenin and the Party leadership, including Nikolai Bukharin, editor of *Pravda*, supported Gastev's approach to work, and the Party placed Leon Trotsky, People's Commissar for Military Affairs, in charge of both munitions manufacture and the railroads. According to Trotsky,

> A whole number of features of militarism . . . blend well with what we call Taylorism . . . If you take militarism, then you will see that in some ways it was always close to Taylorism. Compare the movements of a crowd and of a military unit, one marching in ranks, the other in a disorderly way, and you will see the advantage of an organized military formation . . . And so the positive, creative forces of Taylorism should be used and applied. (Biessinger 1988, 33)

Under Trotsky's aggressive imposition of Taylorite methods instituted largely by Taylorist engineers from the old regime, both the munitions industry and the railroads rapidly improved.

When Trotsky and Gastev's coercive version of Taylorism came under strong attack in 1923 and 1924 as subordinating the individual worker's interests to "the fetish of production" (Bailes 1977, 386). Lenin and the Party leadership supported the advocates of Scientific Management for pragmatic reasons. According to Bukharin, "We must direct our efforts. . . towards creating in the shortest possible time a definite number of qualified, especially disciplined, *living labor machines*, readily available to be put into general circulation" (Bailes 1977, 387, italics mine). Thus an admittedly coercive form of Scientific Management became common in the Soviet Union, initially to meet critical wartime conditions. But the approach was maintained well beyond the crisis stage, usually linked to the problematic claim, initiated by those who knew Taylor almost exclusively through *The Principles of Scientific Management*, that worker dissatisfaction could be controlled almost completely by raising real wages.

Given Gastev's positive attitude toward engineers and his great enthusiasm for Scientific Management as well as for the Bolshevik party, there was one major problem. Prior to the revolution, Taylorism had been promoted in Russia largely by engineers who were supporters of the anti-Bolshevik Provisional Government during 1917. In 1919 at the Eighth Congress of the Communist Party Lenin had insisted that for practical purposes the Bolsheviks would have to use these "bourgeois engineers," and he assured party members that when they observed the successes of socialism those engineers would "themselves be attracted to our *apparat* and make themselves a part of it" (Bailes 1978, 51). But Lenin's prediction was slow to come true. As late as 1928, only 138 out of about 15,000 Russian engineers were Communist Party members.

In 1922, the All Russian Association of Engineers had explicitly distinguished itself from the Party's All Russian Union of Engineers in a way that perfectly reflected the "apolitical" stances of American scientific managers and technocrats:

> So far as the guiding ideas and catchwords [of the Association] are concerned, the first of these is "apolitical"—being apolitical in the sense that the Association of Engineers does not adhere to the ideology or tactics of any political party. But this does not mean that the organized engineering community should not have its own opinion on . . . economic questions and technical questions of wide social import. (Bailes 1974, 453)

Because the number of "Red Experts" was so small, it was inevitable that both Gastev's Central Institute of Labor and the more policy-oriented and important Scientific-Technical Administration of the Supreme Council of the National Economy were staffed overwhelmingly by non-communist members of the All Russian Association of Engineers who continued to argue in 1929 that "the future belongs to managing engineers and engineering managers" without reference to either Marxism or the proletariat (Bailes 1974, 457).

To contradict the independent engineers' claims, Joseph Stalin, who succeeded Lenin as leader of the Soviet state, had begun to gradually replace bourgeois engineers with Communist managers in industrial facilities as early as 1926. October of 1929 saw the beginnings of Stalinist attacks on "The pretensions of the technical intelligentsia and the theses of the scientific-technical administration." The attacks continued with a series of show trials of engineers who had been leaders of the All Russian Association of Engineers, critics of the first five year plan in 1929, and spokesmen for the primacy of technocratic management-engineers.

During the 1930s, while the "old" engineers were purged, young technical graduates, usually drawn from the peasantry or working classes, flocked to the Communist Party, so that by 1937, the 138 engineering members of the party in 1928 had become 47,000. "Red experts" began to replace both old engineers and experienced workers who had come up from the shop floor as managers in Soviet industries. Understandable resistence on the part of experienced skilled workers to being made subordinate to younger "experts," who were often less knowledgeable about specific specialized practices, was fought largely by increasing factory discipline and giving the new managers power to involuntarily transfer, discharge, and even imprison workers who did not cooperate with their leadership. Arriving as little as twenty minutes late could lead to transfer to forced labor with no monetary compensation.

In 1935 a new movement encouraged by Stalin—the Stakhanovite movement—so named to celebrate the accomplishments of a coal miner who mined fourteen times the normal amount of coal in a single shift—began to develop. This movement promoted the special technical education of promising workers so that they could become management experts, with the long-term expectation that the general level of workers would be raised until all management came from the proletariat (which was an expectation of the original Scientific Management movement). But the Stakhanovites also used some of the most unfair practices associated with Taylorism to speed up work processes, sometimes with disastrous consequences.

In order to make a living wage based on the piece-work payment schedules, workers often ran machines constantly without taking time to fill their oil cups, or they tried to get machines to take deeper cuts than they could without doing damage to the machines. According to American and other foreign advisors, the losses in machine damage was often "greater than the value of factory output" (Merkle 1980, 126). Thus, as proletarian management replaced bourgeois management, coercive techniques drawn from Taylorite Scientific Management were often retained even though they were ineffective.

Their commitment to various forms of Scientific Management was only one aspect of a broader technocratic perspective brought to the Soviet Union by its first three Communist leaders, Lenin, Stalin, and Nikita Khrushchev, even though none had been trained as engineers. For them, as for many Marxists including Marx himself, while class conflict was, in one sense, the driving force of history prior to the triumph of the proletariat, ultimately it was changes in the means of production brought about by technological innovations which reshaped the relations of production and thus the relationship between classes, and it was technological change that would continue to produce progress even in a classless society. This version of Marxism, according to which "machines make history" (Josephson 1995, 522), was wide spread among Bolsheviks, and all three initial Soviet heads of state shared a strong sense that science and technology could solve virtually all social problems. Moreover, all showed a practical preference for leadership by a party elite which disdained democratic politics.

With Lenin, the belief in the social power of science and technology was symbolized in his maxim that "Communism equals Soviet power plus electrification of the entire country." According to Paul Josephson,

> Lenin believed electrification would change the shop floor into "clean bright laboratories worthy of human beings," and that household electric light would ease the life of "domestic slaves." . . . [He] believed electricity would lead to agricultural modernization, . . . and eradicate the cultural and economic chasm between town and country, transforming peasants "into full members of the socialist state." (1995, 524)

Stalin's confidence in the transformative power of science and technology was manifested in his construction of the magnificent Moscow subway, the initiation of multiple projects to build canals and hydroelectric generating plants, and in his disastrous support for T.D. Lysenko's Lamarckian attempt to produce new strains of wheat that would withstand harsh weather (Medvedev 1969). Under Khrushchev, in addition to large-scale projects in agriculture and the chemical industry, major efforts in nuclear research and the space technology which eventuated in the orbital flight of Sputnik were undertaken.

Well before John F. Kennedy initiated the Apollo project to place a man on the moon as a technological display of the superiority of American culture and governance, Soviet leaders consciously promoted large-scale technological display projects as symbols of the superiority of Soviet socialism. Gigantic projects such as the Dneprostroi power station, the Magnitogorsk steel mill, the Gorky automobile factory, and the Moscow subway (Rassweiler 1988) were modeled on the latest and largest American or European facilities and then planned to be bigger and better. Lazar Kaganovich, who Stalin placed in charge of the construction of the Moscow subway, for example, wrote in a spirit worthy of Gastev that the Moscow subway,

> far exceeds the bounds of the usual impressions of technological construction. Our metro is a symbol of the new society that is being built. . . The metro isn't simply marble . . .no! It isn't simply granite. . . no! It isn't simply metal . . . no! It isn't simply concrete . . . no! In each piece of marble, in each piece of metal and concrete, into each step of the escalator penetrates the new spirit of man,

> our socialist labor, there lie our blood, our love, our struggle for a new man, for a socialist society. (Josephson 1995, 532)

Huge canal projects to provide both power and irrigation were initiated. Just between 1952 and 1956, for instance, some 110,000 square miles of irrigated land were added to Soviet agricultural land.

But in the absence of any effective political opposition to suggest alternative uses of the vast resources poured into these structures or to express concerns regarding the safety or environmental impacts of such gigantic projects, the costs were huge. Engineers were pressed to complete projects in record time with poorly trained laborers, often with slave labor. Resources were often robbed from food production or from consumer-oriented light industry to focus on these massive show projects. Over 70,000 workers and 5000 engineers, for example, were transferred to the Moscow subway project (Josephson 1995, 541). Ill-trained operators often inadvertently destroyed the derricks and cranes they were working with; on the other hand, laborers were often so harshly treated that they died by the tens of thousands. Canal, dam, and irrigation projects often had disastrous environmental consequences, "they increased evaporation, salinity, and pollution of water resources and destroyed productive farm land, submerging it underwater. In all, 2,600 villages, 165 cities, almost 30,000 square miles (the area of Maryland, Deleware, Massachusetts, and New Jersey combined), including nearly 12,000 of agricultural and 12,000 of forest land, were flooded by Soviet dams" (Josephson 1995, 536).

The Soviet Union expanded its interest in atomic energy from military to civilian uses relatively early and placed the first nuclear reactor on its electrical grid in 1954 before any other nation. And the Soviet peaceful atom project showed virtually all of the characteristics of earlier technological display programs, focusing on the rapid deployment of standardized technologies with relatively little concern for safety because there was virtually no accountability to an informed and risk aversive public. Because engineers were confident about the reliability and safety of their reactor designs, reactors were built without containment vessels and with few redundant safety systems. To make them cheap to build, designs were standardized without regard to special problems of siting. And to make them economical to run, non-technical staff were trained just enough to keep them running in non-crisis situations. Under such circumstances disasters such as that at Chernobyl were almost a certainty in the long run (Josephson 1986).

Sweden

As in France and Germany, initial interest in Scientific Management emerged in Sweden among engineers who were first impressed with Taylor-White Steel, but unlike those cases this initial interest had little extended impact (Runneby 1978, 34, n46). More extensive interest arose as a consequence of a parliamentary committee set up to understand why so many Swedish working men were emigrating to the United States at the beginning of the twentieth century and to suggest how Swedish industries might successfully compete with those in America not only for skilled workers, but for customers as well. When interest was revived, Scientific Management, business management, public administration, architectural modernism, and more broadly conceived "social engineering" gradually came to pervade the entire national culture.

In preparation for its Emigration Report, published between 1908 and 1913, members of the parliamentary committee carried out a number of study trips to the United States in order to investigate, "the productive capacity of American industry, management questions, work intensity, and class reconciliation" (Runneby 1978, 22). These study trips brought Swedish industrialists and business leaders into contact with advocates of systematic management even before it became Scientific Management and before Taylor published *The Principles of Scientific Management*—which was published in Swedish by The Swedish Industries Association in 1913.

Between 1890 and 1910 Sweden had begun a period of rapid industrialization during which annual GDP per capita grew even faster than that in North America (2.4 percent vs 2.0 percent) (Schön 2012, 137), but Swedish employers were convinced that Sweden needed "a harder work effort," and like Taylor they insisted that traditionally the workers' sole goal had been "to obtain the highest possible pay for a minimum of work." In a speech to the Chamber of Commerce of Gäyle in 1911, the economist Eli Heckscher expanded the emphasis on an improved work ethic to include all Swedes: "If the whole people will not make the utmost effort, we cannot raise our economic level—the whole people, not only the workers (Runneby 1978, 23). Among those who embraced Scientific Management early and promoted its virtues, three of the most important were Gustav Aksel Jaederholm, Erik Forsberg, and a Finn, Jakob Johannes Sederholm, who published in Swedish as well as Finnish. All three shared Hecksher's view that efficiency was needed everywhere, not just on the factory floor, and all tempered some of the harshest methods suggested by Taylor by appropriating elements from Hugo Munstererg's *Psychology and Industrial Efficiency*—which was translated into Swedish and published by the Swedish Industries Association in 1917.

Jaederholm was a psychologist who had previously complained about the peer-pressure experienced by workmen to keep the pace of work slow when the editors of *Industria*, the journal of the Swedish Employer's Association, invited him to write several articles about American wage incentive systems and Scientific Management. He promoted Taylor's system as "the first fully developed, modern wage technique"; he endorsed the time and motion studies of the Gilbreths; he insisted on using scientific methods to place workers in positions suited to their aptitudes; he praised the extension of efficiencies to be gained through Scientific Management to all forms of human work; and like Taylor, he argued that for workers, "the fundamental driving force must be *discipline*" (Runneby 1978, 36). Perhaps most importantly, Jaederholm argued that "Democracy" was "a method to reach efficiency." But like Henry Gantt, he defined Democracy not in terms of the establishment of policies by those who were affected by them, but in terms of allowing policies to be set by those who were most capable and knowledgeable about the issues involved, much as the policies of the Prussian Army were established by trained military experts (Runneby 1978, 35–36).

This understanding of "Democracy" became widely adopted within the Swedish population, which became increasingly inclined to understand social policy making as "social engineering" by a technocratic educated elite applying objective knowledge. Notions of scientific management quickly migrated out of the realm of capitalist businessmen into the increasingly important domain of the left-leaning Social Democrats, whose political party became the long-term reigning party in Sweden in 1932. According to Alva Myrdal, one of Swedish Social Democracy's major ideologists, the social engineers "now had to take the step from the plain registration of facts and analysis of causal connections to setting up rational plans for appropriate changes" (Hirdman 1994, 76). And according to Pierre Guillet de Monthoux, the role of the Swedish state "is to function as a wall, from which the balls of conflict bounce back into the realm of managerial problem solving" (1991, 28). Taking their cues from the German rationalizers after von Moellendorf, Swedish Social Democrats saw Taylorite techniques under socialism as serving the interests of the working class by offering more goods at cheaper prices while allowing wages to rise, and Swedish Trade Unions gradually accepted that notion much as those in Germany had.

Forsberg had been trained as an engineer, and like the American Taylorites, he saw engineers as capable of eliminating the traditional antagonisms between workers and employers through the use of "technical improvements, good organization, and good management" (Runneby 1978, 37). Aware that Taylor's system had not been widely adopted in the United States, Forsberg was convinced that its failures had been produced because it had been slandered and misinterpreted. He attempted to institute Scientific Management at a factory in Olofström where he too met with resistence on the part of both employers and workers. He complained that "We fought not only for our own sake, but also for a vitally important principle, that of Scientific Management, whose victory could have resulted in immense advantages to the whole mechanical industry" (Runneby 1978, 39).

In 1916 Forsberg continued his campaign for Scientific Management in a textbook, *Industriell ekonomi,* which included calculations of production costs, promoted Taylorite techniques of shop management, and gave practical examples of time and motion studies applied to Swedish industries. Along with the Swedish edition of *The Principles of Scientific Management (Rationell arbtsledning)*, Forsberg's text formed the core reading in Eli Hecksher's courses in company organization taught at the Stockholm School of Economics (Engwall and Pahlberg 2001, 16). Many of Hecksher's students traveled to the United States and studied applications of Scientific Management, and two of them, Ruben Andersson and Gerhard Törnqvist, wrote a joint master's thesis, *Scientific Management: A Study of Its Growth and Some Sample[s] of Its Present Practice*. Andersson, after changing his name to Rausing, went on to become founder of the global packing giant, Tetra-Pak, whose practices were grounded in Taylorite ideas, while Törqvist took over lecturing on business economics and Scientific Management at the Stockholm School of Economics in 1922 (Engwall and Pahlberg 2001, 19). These courses and the similar courses of his successors, constituted the only business management education available in Sweden prior to the 1960s, and they became the chief source not only of the managers of Swedish industry and business but also of bureaucrats in the public sector (De Monthoux 1991, 32).

Sederholme was a scientist who was the director of the Geological Commission of Finland, but he had wide-ranging social and political interests. As early as 1904 he was lecturing on the relative efficiency of American industries, emphasizing mechanization, specialization, standardization and mass production, though he was concerned that in America, men might become "slaves to their work" (Runneby 1978, 41–42, n.98). After learning of Frank Gilbreth's writings on brick laying efficiency he traveled to the United States in 1913. At the recommendation of his fellow geologist, Charles Van Hine, who was the president of the University of Wisconsin, a center of Progessive thought on efficiency, Sederholme contacted both Taylor and Morris Cooke, with both of whom he carried on substantial long-term correspondences, and he visited the standard model industries including Link-Belt and the Watertown Arsenal. On returning to Scandinavia, Sederholme began lecturing on the virtues of Scientific Management, especially to trade union groups and university student clubs, trying to spread the gospel of Taylorism beyond the engineering community. In fact, he presented Cooke's Carnegie study on efficiency in higher education to the rector of the University in Helsinki (Runneby 1978, 42–43). In 1916, Sederholme published the first major Scandinavian book devoted solely to Taylorism, *Arbeiets vetenshop (The Principles of Work).* There he argued for the extension of Taylorite practices to business, banking, agriculture, schools, universities, political bodies, household management and one's personal life. In all of these realms Scientific Management could resolve conflicts and create progress. Here, according to Sederholme, "the belief in the ability of science to create happiness appears in a new form" (Runneby 1978, 44).

One of the first areas in which Social Democratic social engineering was widely applied was housing. The Social Democrats saw housing construction as an ideal way to pull Sweden out of the global depression of the early 1930s. But the new form of housing would not be traditional; instead, it would both be efficient on its own and promote the efficient use of social resources. In particular, it would be designed to free women from confinement to the house in order to allow them to enter the workforce. According to Sven Markelius, one of the architects who were influenced by Corbusier and who were close to the Social Democratic intellectual leaders, Alva and Gunnar Myrdal, housing should be designed on the assumption that children would not ordinarily live with their parents, but would be sent off to boarding schools, because "the housing of children during the night in the parents' apartment would mean that in every family, at least one of the parents would be tied to the home during the free hours after a day's work" (Hirdman 1994, 79). The new home would be a "white, flat roofed, functional . . .'house-machine'. . . [divided] according to various functions: the functions of work, of rest, of sexuality: and all this within . . . about 40 square meters for a small family." Thus, writes Yvonne Hirdman, "the small flat with a mini-kitchen, a living room, and—of course—a bathroom, was introduced into Sweden" (Hirdman 1994, 79). Furthermore, by the mid 1950s this home would come to be furnished with functional furniture from Ikea, a company run along Taylorist principles and employing modernist design principles:

> To trim cost, Ikea's design engineers work closely with suppliers, designing furniture from the start for mass production. By dealing directly with factories, the company eliminates middleman costs. It also uses manufacturers that are best equipped to make furniture components. Instead of traditional furniture factories, for example, Ikea relies on a shirt manufacturer to make seat cushions and a window manufacturer for table underframes. (De Monthoux 1991, 31)

The emancipation of women envisioned by Alva Myrdal and her young architect friends did not happen as rapidly as they had hoped, in part because the perception of population decrease caused by emigration led to a new emphasis on women's reproductive and mothering functions. By the mid-1950s, however, new programs, such as the provision of school lunches, made it possible for many women to be employed outside the home. And gender-neutral maternity/paternity leave policies made Sweden one of the first countries to see a significant proportion of families in which traditional gender roles were reversed, with fathers taking over parenting and home-making duties while their wives worked in the "public" economy.

The transformation of Swedish social and political life to one grounded in the concept of management by a technical elite educated in schools of business management has been a mixed blessing according to Pierre De Monthoux. In order to make the "Thought Taylorism," which is taught in the business schools and universities and which is applicable to any kind of human activity, sufficiently scientific and universal, its techniques had to be dissociated from their historical and cultural contexts. De Monthoux argues that,

> this striving to turn an art into a science, to make the special look general, severs the solutions from their contexts and makes nonsense of them. . . . [This has] resulted in a kind of training that is non-theoretical (as in the case of political economy without research contact) and non-historical (as in the case of business economy divorced from business history). What was taught tended to become a form of modern scholasticism or a kind of nonsensical drill. And yet student numbers increased, especially when most Swedish universities also started training managers in the 1960s. (De Monthoux 1991, 34)

Japan

The Japanese appropriation of Scientific Management was at the same time more popular, more widespread, and more deeply modified to fit new cultural circumstances than it was elsewhere. The journalist, Ikeda Tōshirō, was in the United States during the Eastern Rate Case episode. Fascinated with Taylorism, he returned to Japan in late 1911 and published a series of articles on Scientific Management in a newspaper, the *Sakigake shinbun*. The success of these articles led Ikeda to write an introduction to Scientific Management in the form of a story of a young boy named Tarō who battled against "soldiering" and waste. His self-published *Secrets of Elimination of Wasted Work (Mueki no tesū o habuku hiketsu)* was an instant success, selling 1.5 million copies, 20,000 of which were purchased by the head of Mitsubishi Electronics to distribute to his employees and 50,000 of which were distributed by the head of Kawasaki Shipbuilding to his workers (Tsutsui 1998, 18–19). Ikeda drew heavily from the popular Meji "success story" tradition, which in turn shared many features with the Horatio Alger tradition in the United States. In both, impoverished boys rose to lives of security and comfort through hard work, determination, courage, and honesty when their work was recognized and rewarded by a rich patron. Ikeda's story thus ignored concern with the more exploitative elements of Scientific Management and focused on the way that it could harmonize the interests of labor and management.

As we shall see, this focus on harmony and cooperation remained a hallmark of Japanese scientific management. According to Ueno Yoichi, writing in 1928 in a text revised and republished in 1929, 1944, and 1956,

> Cooperation is absolutely necessary in all organizations. In business, success is doubtful without cooperation. If all, from the president on down to the lowest functionary, can pool together their various personal abilities, work together for the common good of the company's prosperity and devote themselves as a group, then great increases in results are possible. (Tsutsui 2001, 455)

In the Japanese context, however, the Taylorite emphasis on cooperation took on a radically different meaning than it had in the United States.

The banker, Hoshino Yukinori, who had also been traveling in the United States during 1911, asked Taylor for permission to translate *The Principles of Scientific Management* into Japanese. Taylor agreed, and the Japanese translation appeared in1913, followed quickly by translations of *Shop Management* and Frank Gilbreth's *Motion Study*.

While early twentieth-century Japanese capitalists could sometimes be as exploitative as those in the West, two factors reinforced one another in leading both Japanese business leaders and Japanese labor leaders to interpret Scientific Management in a unique way. First was a long Meji tradition which emphasized the mutual responsibility of clients to their patrons and of patrons for the welfare of their clients. Within growing Japanese capitalism this meant that many owners and managers accepted an obligation to promote the welfare of their workers. Second, in large part because Scientific Management entered Japan during a period of severe labor unrest, skilled labor shortage, and rapid workforce turnover, company concerns were often heavily focused on retaining capable workers while increasing productivity. In connection with these two conditions, the emphasis on cooperation between labor and capital which had been aimed solely at increasing wages and productivity in the American context was transformed into a tool for promoting a vision of the factory as a moral community in which the workers' service to the managers was rewarded through much broader consideration of worker welfare by management. In return for loyal service to the company they worked for, employees could expect a lifetime guarantee of employment, opportunities for advancement to positions appropriate to their ability levels, and regular pay increases based on seniority. In addition, Japanese managers generally agreed to consult with workers regarding work processes and their impact on worker morale.

According to Uno Riemon, one of the most prolific Japanese writers on Labor-Management relations during the second and third decades of the twentieth century,

> our ideal of the way to treat workers [is] the system of constitutional polity in enterprise management. In brief, under this system both workers and employers participate in decisions concerning matters of importance in managing the factory, and particularly those relating to the workers themselves. It is implemented through the united efforts of the factory owners and the workers. The goals [are]: (1) to cause workers to think of the factory's business as *their own*, and to unite their will with the business; (2) for factory owners and workers to reach full mutual understanding; and (3) to see that all matters are carried out satisfactorily and through mutual cooperation. (Hazama and Kaminski 1979, 98–99)

Thus Uno and others adopted the practices of industrial democracy which were accepted by members of the Taylor Society after 1915, when Taylor died. Companies, including Mitsubishi Electric, which were committed to adopting Scientific Management, established co-operative Workshop Committees including both management and workers to initiate efficiency improvements as well as to consider worker welfare. Such committees tended to reduce worker resistance to the introduction of new techniques. Companies also followed Taylor's recommendation to establish in-house promotion systems to raise both retention and morale (Sasaki 1992, 14–15).

While many features of American Scientific Management could be utilized or modified to fit the Japanese circumstances, Taylor's original focus on rewarding individual effort solely with financial incentives had to be modified for the Japanese situation. According to Uno, applying unmodified American Scientific Management had led to numerous failures, and, "The reason for these failures lies in having brought in American style [practices] which attach too much importance to material things, and using them in an unmodified form on Japanese workers, who are more inclined to be moved by things spiritual" (Hazama and Kaminski 1979, 100).

Some 200 Japanese were granted government fellowships between 1896 and 1916 to study American industry and commerce. A significant number of these ended up working for the Japanese National Railroad system which had nationalized seventeen private railroads during 1906 and 1907. When nationalization occurred, there were some 190 types of locomotives used by the seventeen private railroads, and the coupling systems for different roads were not compatible with one another. Under the leadership of mechanical engineer Yasujiro Shima, who had graduated from the Imperial University in 1894, rationalization and standardization of rolling stock was undertaken. Locomotive varieties were reduced to three, one for passenger trains, one for freight trains, and one for switching purposes. Though it took him a while,

Shima, managed to switch all coupling devises to a single automatic variety on a single day, July 17, 1926. Factories to repair rolling stock were equipped with specialized machines, such as overhead cranes; and the layouts were redesigned to make work flow efficient. Furthermore, under the direction of Kennji Kishiyama, who studied Taylorite methods at a factory of the Chicago and Northwestern Railroad, several efficiency improving methods were implemented at repair facilities.

Preventative maintenance schedules were established, beginning in 1910, allowing for evening out workload. A system for stocking adequate supplies of spare parts was created in 1914–15. A schedule of times for different repair procedures was established. Functional foremanship was introduced around 1922, though wage incentive systems, including the piece rate system based on motion studies were delayed until 1929 in favor of group discussion and problem-solving teams (Daito 1989, 14–19; Strang and Kim 2003, np). As a consequence of these changes, the average number of man-days to repair a locomotive was reduced from 900 in 1912 to less than 350 in 1926, and the average number of days that a locomotive was out of service for repairs was reduced from over forty to about six during the same period (Daito 1989, 11–12).

Private companies in Japan increasingly manufactured rolling stock for the National Railroad. Shima chose four firms as suppliers in 1912—Kisha Seizo, Nippon Sharyo Seizo, Kawasaki Shipyard, and Amano Factory—and in order to keep costs down Shima promoted the use of Taylorite methods by his suppliers—by encouraging JNR trained engineers to transfer to the private suppliers, by providing extensive recommendations on what kinds of specialized machinery to purchase, and by purchasing advanced American and German locomotives and providing them to suppliers to be reverse engineered (Daito 1989, 22). Several private companies not related to the National Railroad also sponsored visits to American companies that were using Taylorite methods. Thus, for example, in 1923 Mitsubishi Electric sent sixty engineers to Westinghouse to bring back information about time studies, process control and cost accounting. Many of these engineers introduced Taylorite methods into shop practices, creating a professional engineering branch of Taylorism in Japan (Sasaki 1992, 16).

Far more important than the engineers who adapted Scientific Management to local industrial practices, however, were a group of businessmen, academics, management consultants, and psychologists, who embraced a more expansive vision of Scientific Management which sought to extend efficiency beyond the factory to commerce in general as well as to the activities of daily life and to the formulation and implementation of government policies. Though often derided by the technically oriented engineers as mere "efficiency peddlers" (Tsutsui 2001, 448), it was largely through their efforts that, "the concepts of Scientific Management spread more widely, penetrated more deeply and remained relevant much longer in Japan than most observers have assumed" (Tsutsui 2001, 444). And it was largely through this second group that Taylor's individualism and disdain for general worker welfare was rejected in favor of viewing a productive enterprise as a community in which "group goals and tasks . . . are held to have priority over individual ones; . . . Work groups have more autonomy as to how tasks are carried out vis-á-vis individual accomplishments, [and] more information is shared with the workforce by managers, who place a premium on competence" (Warner 1994, 510).

The most important representative of this second group was undoubtedly Ueno Yōichi (1883–1957), who obtained a degree in psychology from Tokyo University in 1908, after which he wrote popular works on psychology and edited the journal *Shinri kenkyū* (Psychological Research), through which he introduced the ideas of both Alfred Binet and Sigmund Freud to Japanese audiences (Tsutsui 2001, 444). Ueno first came into contact with Scientific Management in 1910 through the work of Frank Gilbreth, but he soon absorbed the entire Taylorite corpus and began to popularize the works of U.S. efficiency experts for the general public. In 1913 he edited a special issue of *Jitsugyōkai* (Business World) on Scientific Management, and as an outgrowth of his lecturing on the subject in 1919, he was hired by Lion Toothpowder Company as a consultant to improve their production practices and advertising, thus initiating a lucrative sideline. Two years later he was recruited by a recently formed semi-governmental think tank, *Kyōchōkai,* tasked with promoting harmonious cooperation in industrial relations. Ueno established a bureau within the organization to deal with efficiency and immediately took the opportunity to travel to the United States, where he toured Link-belt Corporation, attended a meeting of the Taylor Society, and met with the Gilbreths and Morris Cooke, among other leaders of the Scientific Management movement (Tsutsui 2001, 447).

In part because of his own background in psychology, and in part because of the Japanese emphasis on worker satisfaction and worker welfare, Ueno focused on deploying Taylorism more in a human relations framework than it was in America. When he organized the efficiency bureau within the *Kyōchōkai*, for example, he included among its functions publishing reports on psychological testing, developing and disseminating aptitude tests, and training industrial psychologists. More importantly, like many Taylorites, and even more like von Moellendorf, with his broader emphasis on rationalization and central control over the economy, Ueno extended his opposition to competition and his focus on cooperation to macroeconomic issues. In an economy subject to free market conditions, "It is inevitable that competition should exist between individual economic units. As a result, however, it reaches the point where one unit gives no regard to the interest of others and, paying no attention either to the movements of the entire national economy, each economic entity just serves its own profits," he wrote (Tsutsui 2001, 460–461). Ultimately, according to Ueno, the principles of Scientific Management should be extended from the shop floor to centrally controlled national economies, and beyond: "The principles which Taylor attempted to establish at one small factory many years ago must eventually develop into [a form of] Scientific Management that encompasses all aspects of human life and seeks to benefit all peoples of the world" (2001, 462).

Searching for a model planned economy, though he praised Lenin's ideals, he preferred carefully regulated capitalism. While there were important elements of von Moellendorf's collective economy which he admired, Ueno felt that it failed because it was not grounded in the prior application of Scientific Management to each productive unit. Thus he wrote:

> The combination of horizontal and vertical combination was the basis of the industrial rationalization promoted in Germany after World War I. However, since this industrial rationalization was not preceded by the establishment of Scientific Management in individual firms, its results were few. Just rationalizing industries without reforming companies one by one . . . is nothing more than substanceless economic concentration. (Tsutsui 2001, 461)

Surprisingly, because he imagined that in the United States virtually all firms had adopted Scientific Management, and because he saw the efforts of both Herbert Hoover and the early New Deal as imposing central controls over the economy, Ueno hailed the United States as a nation moving inexorably toward the kind of planned economy which he advocated (Tsutsui 2001, 462).

When the process of rationalization according to the tenets of Scientific Management was completed, Ueno offered the following Utopian vision:

> If the rationalization of the citizens' lives described above occurs, then the nation will literally become an extended family, the requirements of life will be guaranteed with the least necessary labor, all the citizens will delight in the creative life and work for the elevation of culture. This will certainly be the realization of a democratic paradise. (Tsutsui 2001, 463)

Of course, Ueno's understanding of democracy was that associated with that of American Taylorites such as Morris Cooke and Henry Gantt. Ueno explicitly rejected the "debating chamber" theory of democracy and denounced politicians as "self-serving, venal vote grubbers" (Tsutsui 2001, 463). Instead he offered a theory of democracy defined by governance by wise men guided by the scientific dictates of efficiency (Tsutsui 2001, 464).

During the period from 1927 to 1937 a new term, "rationalization," (*sangyō gōrika*) replaced that of "efficiency"(*nōritsu*) as dominant in Japanese discussions of economic improvement. Japanese rationalization differed from its German inspiration by following Ueno's advice to incorporate Scientific Management concepts both at the level of the individual firm and at the level of centralized economic management. Though there were many definitions of *sangyō gōrika*, all, according to William Tsutsui, included the following elements: First, *sangyō gōrika* depended on the replacement of tradition and rule of thumb with the rule of "science" and of fact. Second, *sangyō gōrika* was opposed to "favoritism, custom, and self-centeredness," or "economic individualism." Third, although there was some disagreement over the role of the state in *sangyō gōrika*, that disagreement was largely over whether the state should merely encourage private industries to standardize products and eliminate waste as was happening in the United States or whether it should dictate standards for both products and processes. Fourth, at the macro-level *sangyō gōrika* promoted both the horizontal and vertical integration of industries as was happening in Germany, and at the micro-level it promoted the incorporation of the latest management techniques and the incorporation of mechanical assembly lines as the last word in Taylorite shop-level attempts to increase efficiency (Tsutsui 1998, 63–66).

As the pressure for an authoritarian state grew through the late 1930s and early 1940s, leading to the New Order Movement and de-facto Imperial control over all economic assets, European fascist theories played a major role. But Taylorite and *sangyō gōrika* residues were extremely important. When the Japan Efficiency Association was created by the government in 1942 as a high-profile organization to improve Japanese industry, the long-time advocate of Scientific Management, Godō Taku was appointed chairman. In his opening address he argued that, "There have long been calls for increased industrial efficiency in Japan, but it has never been sought so insistently by the state nor been expressed with such newfound significance." Then he went on to identify three principles to guide the Association: (1) "[to] devise and implement efficiency methods imbued with the Japanese spirit. (2) [to] emphasize practice over theory, and (3) [to] dispense with equality" (Tsutsui 1998, 94–95)—all principles thoroughly consistent with scientistic and technocratic impulses.

China before the People's Republic

The Chinese case offers a slightly different pattern of transmission than those we have seen elsewhere. Taylorism arrived in China largely at the hands of Chinese students who had studied at universities in the United States. Yang Xingfo studied in the United States between 1912 and 1918, completing an engineering degree at Cornell and an MBA at Harvard. While at Cornell he happened to take the first course on Scientific Management at any university, given by the mechanical engineer Dexter Kimball. In 1914, Yang, who went on to become the secretary general of *Academia Sinica* in 1928, was leader of a group of Chinese students in America who established the Science Society of China, whose aim was "the diffusion of the scientific spirit and knowledge for scientific growth and industrial prosperity" (Kwok 1971, 14), and whose journal, *Kexue (Science)*, was first published in the United States but moved to China in 1918. For one of its first issues Yang wrote an essay, *renshi zhi xiaolu* (personal efficiency) based on Taylor's ideas.

More important initially was Mu Xiangyu, member of an important cotton mill owning family. In 1909, at age thirty-three, Mu came to the United States, getting a BS in agriculture at the University of Illinois and an MA in agriculture at Texas A&M. While at A&M he was introduced to Taylorism, and he contacted Taylor asking permission to translate *The Principles of Scientific Management* into Chinese. Taylor was very enthusiastic, even sending Mu a copy of Hoshino's 1912 translation into Japanese, which Mu used to guide his translating effort (Morgan 2006, 411), completed with the help of Dong Dongsu in 1916. Back in China, Mu set up the *Deda* Cotton Mill in Shanghai, the *Housheng* cotton mill, and the *Zhengzhou Yufeng* cotton mill, using Taylorite criteria to control the quality of the contract labor that was ubiquitous in China at the time, and providing detailed written operating instructions and disciplinary rules, all of which led to improved quality and productivity (Morgan 2006, 412). He went on to establish the Chinese Cloth and Yarn Exchange in Shanghai, finishing his career in government service as deputy minister of industry during the 1930s.

Interest in Scientific Management surged in China in the 1920s, when European companies, recovered from WWI, began competing with Chinese firms at the same time that workers were demanding higher wages and better working conditions. Several enterprises initiated Taylor-inspired methods to increase productivity and to justify higher wages. Among these, the Kanguan Can Factory and the Commercial Press became the Chinese equivalents of Link-Belt and the Tabor Manufacturing Co. Commercial press, for example, was able to raise productivity by 250 percent while increasing wages by 20–30 percent and greatly improving employee discipline. During the late 1920s the popular journal, *Dongfang zazhi (Eastern Miscellany)* began to run substantial numbers of scientific management-related articles, including articles on the German rationalization movement. Then in 1930, prompted by the International Management Institute in Geneva, the Chinese Nationalist government promoted the establishment of the Chinese Industry and Commerce Management Association (CICMA) by a group of Chinese industrialists. The board of directors included both Mu Xianyu and Yang Xingfo as well as a group of very prominent businessmen, and it was chaired by Kong Xiangxi, who pushed the group to promote Scientific Management as a way to improve production skills, reduce waste, and foster the growth of industry for the benefit of the masses (Morgan 2006, 413).

Members of CICMA who attended a November 1930 National Industry and Commerce Congress pushed through a resolution asking the government to direct all provincial and municipal government agencies responsible for industry and commerce to implement Scientific Management methods, to set work performance standards, and to make certain that workers who exceeded the standards were rewarded. In January of 1931 the Ministry of Industry at least nominally implemented the Congress's recommendations. Moreover, CICMA used the Ministry of Industry's journal to promote Scientific Management through 1934, when it established its own journal. While articles on production management and financial management appeared in substantial numbers, the greatest concerns of Chinese managers were with personnel management—eliminating contract labor, finding, training, and then retaining good workers by providing health care, insurance, and pension schemes.

Long-standing Chinese practices often made a complete movement to Scientific Management impossible. Thus, for example, Rong Zongjing, a member of the CICMA board, continued to recruit senior staff for the family enterprises based on family connections or place of birth, rather than technical competence. But the elimination of contract labor and the efficient organization of shop floor activity allowed for the reduction of 50 percent of the workers at the *Shenxin* Cotton Mills with no reduction in productivity (Morgan 2006, 417).

Some of the technocratic elements of Taylorism penetrated the Nationalist Chinese government's scientific management efforts—evidenced by the expert management of government-operated enterprises like the Chinese railroads, which were extremely efficient—but Chinese managers never had to face a well-entrenched Parliamentary democracy as in France or a representative democracy as in the United States, so they rarely had to articulate any explicit opposition to "political" decision making. Nor was there a tradition of complaining about those above one in a hierarchy, even if that person was technically inferior, so there were few, if any, public complaints that decisions made by family members were inferior to those that might have been made by subordinate engineers, whatever the engineers' feelings might have been, although such a notion was implicit in the claim that scientific managers could mediate between ownership and labor.

An Important Qualification

The Scientific Management-based brand of management was successful almost everywhere outside of the United States both in firms and in the domain of social policy making within its context of origin, which allowed for the resolution of earlier tensions between employers and workers. But within the new context of lean production, which is coming to dominate global commodity production, flexibility and profitability have been emphasized at the cost of long-term planning, worker protection, and employment security, undermining the cooperation between worker and employer. Moreover, the need to compete globally has driven both profits and wages down, making it increasingly difficult for any nation to fund the kind of social welfare programs which defined Swedish Social Democracy from the 1930s through the 1990s at least. These changes have led to the feeling that lean production is leading to "mean production" and a turn to a neo-Taylorism which features the coercive elements of Taylorism that dominated the Soviet Union during the 1920s and 1930s rather than the cooperative form of Taylorism represented by the workplace democracy movement which engaged Scientific Management experts in the United States after 1915, Social Democratic policies in Sweden from the 1930s to the beginning of the twenty-first century, and the long-standing Japanese pattern of harmony and cooperation (Nilsson, 1996, 450). The implicit promise of Scientific Management was that if a worker met "scientific" standards of productivity she was guaranteed employment, but on-demand production makes that promise difficult for managers and owners to keep.

Chapter Four

Technocracy during the Cold War, 1945–1990

The production demands of WWI had promoted the use of Scientific Management among virtually all Western industrialized nations as well as the Soviet Union. Moreover, the success of the new techniques for improving productivity promoted the extension of many of those techniques beyond industrial applications to broader commercial and governmental enterprises. In addition, WWI was the first war in which science-based technological innovations played a major role. The use of chemical warfare, in particular, signaled a change in the character of military operations, and Fritz Haber's process for producing ammonia radically changed the manufacture of explosives, leading some to identify WWI as "the chemists war" (Freemantle 2014). So when WWII began there was an expectation that science-based technological innovation would once again play a significant role. The German deployment of bomb-carrying rockets and the Allied development of radar and the atomic bomb as well as the development of the new mathematical techniques of Operations Research for guiding miliary tactics were among the most obvious proofs that those expectations had been realized.

At the end of the war it became clear very rapidly that a realignment had occurred which left two armed camps facing one another in a state of tension. The Soviet Union, led by Joseph Stalin, and its communist block allies stood against the United States and its allies, who formed the North Atlantic Treaty Organization in 1949 to ensure mutual support.

The Soviet Union after Stalin

Stalin's death in 1953, and even more importantly the critical impact of scientific and technological innovation during WWII, effectively revived the central role of highly educated technical experts in the Soviet Union. Abandoning Stakhanovite ideals, highly educated "Red experts" who had become party members in the thirties moved into political leadership positions (Beissinger 1988, 7). Nikita Khruschev was the last major Soviet leader to come from the proletariat without getting a college degree. Leonid Brezhnev, who succeeded Khruschev as General Secretary of the Communist Party from 1964 to 1982, for example, trained as a metallurgical engineer and worked in the iron and steel industry in the Ukraine before taking up a political career. His successor, Yuri Andropov, received his degree from the Rybinsk Water Transport Technical College. Vasili Kusnetzov, who was Deputy Secretary of the Supreme Soviet from 1964 to 1984 and acting Secretary during three different regime changes, had studied engineering in Russia and metal processing in the United States from 1931–1933. Alexi Kosygin, chairman of the Council of Ministers from 1964 to 1980 was a graduate of the Leningrad Textile Institute and had been director of a textile mill and head of Soviet light industry, and Andrei Gromyko, Chairman of the Supreme Soviet from 1985 to 1988, studied at Borisov technical school before going to Moscow University where he got an economics degree. Not until 1988, when Mikhail Gorbachev became Chairman of the Supreme Soviet, was a lawyer rather than a technocrat the Soviet head of State.

Not only were Brezhnev and Kosygin technocrats in terms of their educations, they also continued the technocratic policies of Stalin and Khrushchev, especially with respect to technological displays, regardless of their negative consequences for the overall economy. Brezhnev, for example, planned to computerize all economic units in the country and link them to a centralized mainframe in order to "perfect" the economic planning system; he pushed to complete a modern trans-Siberian railroad (BAM) to promote the extraction of Siberian resources without regard for the financial costs or the destruction of forests and creation of erosion; and he authorized the construction of two paper mills on the shores of Lake Baikal, finished in 1967 and 1973, both of which release masses of untreated heavy metals, sulfur, and other pollutants into the lake (Josephson 1995, 551–555).

A critical consequence of the dramatic Soviet technical achievements associated with WWII and the Cold War, which Soviet scholars identified as the "Scientific-Technological Revolution" (STR), was that Soviet ideology evolved during the 1960s and early 1970s to place scientific and technological change rather than class conflict at the center of social transformations (Aronova 2011, 188). This idea was not taken as a replacement for Marxism-Leninism, but rather as a reinterpretation of Marxism-Leninism for the modern world, and it was intended to guide practical politics as well. According to D.M. Gvishiani, deputy head of the State Committee on Science and Technology, as well at Kosygin's son-in-law,

> This interest is by no means of an academic nature. The communist and worker's parties are elaborating a theory which is a guide to action and forms the scientific basis of their practical program. Speeding up scientific and technological progress to the utmost and making use of achievements of the scientific and technological revolution are the main ways of attaining further development of the material and technological basis of socialist society and its growth into the material and technological basis of communism (Hoffman & Laird 1985, 2).

Furthermore, the STR brought back the notion that scientific management (now much more broadly construed than Taylorism) and scientific Marxism were very closely related and that scientific management should be applied to all levels of organization, including the government and society as a whole. While it is probably true that the development of some sophisticated mathematical approaches to policy making, including game theory, grew more slowly in the Soviet Union and were less frequently applied to policy issues than they were in the United States, the goal of establishing scientific ways to determine a broad range of policies was pursued and was focused largely on Cybernetics—the science of feedback and control (Erickson et al. 2013, 17–21).

V. G. Afanasyev and the Scientific Management of Society

Beginning in the mid-1960s, hundreds of articles and books on the new principles of scientific management, focusing on the need to increase efficiency and productivity (the original goals of the old Scientific Management), appeared in the Soviet Union. Moreover, between 1965 and 1968, more than thirty separate conferences were held on scientific management research (Beissinger 1988, 180). Though many scholars were involved in promoting this new version of Marxist-Leninist scientific management, one of the most important in terms of communicating the professional scholars' ideas to a very broad audience was V.G. Afanasyev, a philosopher/journalist, who spent much of his early career teaching the principles of Marxism-Leninism at the Chelyabinsk Pedagogical Institute. His authorship of the multivolume and frequently republished *Marxist Philosophy* got him an appointment to the philosophical branch of the Academy of the Social Sciences of the Central Committee of the Communist Party of the Soviet Union and a deputy editorship of *Pravda* beginning in 1968.

In 1971, Afanasyev's *The Scientific Management of Society*, which offers an excellent and readable history of management theories applied in both capitalist and socialist societies from the appearance of Taylor's *Principals of Scientific Management* to Soviet works written as late as 1968, appeared in an English version. In the final chapters of this work Afanasyev clearly promoted the latest techniques in management, including mathematical model building, for use in the scientific management of society which, he argued, could be characteristic of the Soviet Union, but was impossible in capitalist countries where centralized planning was impossible (Afanasyev 1971, 273). Moreover, in his subsequent *The Scientific and Technical Revolution—Its Impact on Management and Education,* Afanasyev summarized the aims of the modern management writers, saying,

> In building Communism Soviet society has entered a period when intensification has become necessary all along the line, when economic growth and socio-political and intellectual development for which it provides the basis are achieved not so much by raising production capacities and numerical increase in the labor force, as by increasing the effectiveness of production and the productivity of labor. (Afanasyev 1975, 317)

Afanasyev was elected to the Central Committee of the Communist Party of the Soviet Union in 1976, indicating that these views were consistent with the views of the Party leadership of Brezhnev and Kosygin. Moreover he was made editor in chief of *Pravda* from 1975 until 1989, when his criticisms of Gorbachev and Boris Yeltsin led to his dismissal.

In practice, Soviet state administration never reached the cohesion projected by scholars such as Afanasyev, in large part because the feedback required by modern management theories was very weak in the Soviet Union and because in the absence of competition the managers of monopolistic industries had little incentive for improving products or production practices (Hoffman & Laird 1985, 63–64). Under Brezhnev and Kosygin, efforts were made to improve the situation with little more than marginal success, and with Gorbachev's rise to power both technocratic emphases and the Soviet Union itself collapsed.

Scientism and Technocracy in the United States after WWII

During WWII and its Cold War aftermath technocratic elements became increasingly important within the United States, especially at the federal level and in connection with issues of defense and international relations. They gradually spread into internal civilian political policies as well during the Kennedy and Johnson administrations, though their impact on non-military issues was substantially reduced under Richard Nixon.

As was the case in the Soviet Union, it seemed clear to many American political leaders and analysts that scientific and technological innovation had played a huge role in the prosecution of World War II. Radar, jet-propelled aircraft, penicillin, and above all, the atomic bomb were the most dramatic symbols of the science-based technologies that shaped the war; but in a less visible way, virtually every weapons system, military strategy, and health measure produced during the war had been dependent on technological innovation. Moreover, as in the Soviet Union, it seemed clear to many in the United States that postwar developments connected with both military and civilian concerns were likely to continue to be driven by scientific and technological considerations.

At the request of President Franklin Roosevelt, Vannevar Bush, an electrical engineer from MIT who had served as head of the U.S. Office of Scientific Research and Development during World War II, produced a plan for governing and promoting scientific and technological research and development after the war. While few of its specific recommendations were ultimately embraced, Bush's *Science: The Endless Frontier* articulated an ideology which underlay the relationship between science and technology (technoscience according to some scholars) on the one hand and the federal government on the other at least into the 1970s and for most purposes at least until 1993.

Five fundamental assumptions which were embedded in Bush's rhetoric formed the basis of most political debates relating to science and technology between 1945 and the last decade of the twentieth century. Daniel Sarewitz has characterized these assumptions as myths which guided the actions of scientists and policy makers alike. They were:

1. The myth of infinite benefit: More science and more technology will lead to more public good.
2. The myth of unfettered research: Any scientifically reasonable line of research into fundamental natural processes is as likely to yield societal benefits as any other.
3. The myth of accountability: Peer review, reproducibility of results, and other controls on the quality of scientific research embody the principle ethical responsibilities of the research system.
4. The myth of authoritativeness: Scientific information provides an objective basis for resolving political disputes.
5. The myth of the endless frontier: New knowledge generated at the frontiers of science is seperable from its moral and practical consequences in society. (Sarewitz 1996, 10–11)

Though only numbers 1 and 4, the myth of infinite benefit and the myth of authoritativeness, are at the heart of scientistic and technocratic thought, numbers 2, 3, and 5 distinguish American scientistic and technocratic attitudes from those dominant in the Soviet Union. While both Soviet and American politicians, as well as scientists and engineers, saw the events of the war years as evidence that science was increasingly important for technical and therefore for social progress. Marxist philosophy had always insisted that science was justified only to the extent that it served social ends. Marxist politicians had always supported almost exclusively mission-oriented sciences, and most, though not all, Soviet scientists were comfortable directing their work toward socially defined goals. In the United States, on the other hand, prior to the war American scientists somewhat disingenuously argued that they had received neither much support nor much direction from the government. Significant numbers did work for government bureaus and others worked in corporate laboratories doing work directed toward corporate goals; but more than half of all American scientists worked in academic institutions and chose their own problems to work on. They had chafed under the restrictions imposed on military research during the war and argued strongly that scientists best understood what problems would yield results when studied—whether or not those problems were immediately applicable to society's needs (Aronova 2012, 319). In the long run they were convinced that new knowledge would yield social progress; but without offering much compelling evidence, they insisted that unfettered research was likely to be more fruitful than externally directed research.

American scientists did want to take credit for the long-term positive consequences of scientific research so they could have a call on society's resources; but they wanted to distance themselves from the negative moral taint associated with the atomic bomb and with chemical and biological warfare. Thus they insisted that technical knowledge was morally neutral and that whatever blame was associated with new technical knowledge should fall not on the knowledge itself nor on those who produced it, but rather on those who used that knowledge for evil purposes. Though not an American, Michael Polanyi, a Hungarian émigré to Britain, and a "philosophical chemist" became one of the chief spokesmen for the view that the factual and calm assessment of situations could transcend ideological orientations and, indeed, offer an "end to ideology" as well as a firm basis for a wide range of political decisions (Aronova 2012, 311–312).

RAND Corporation and the Rise of Think Tank Expertise

Virtually all of the assumptions associated with Bush's *Science: The Endless Frontier* and the views of those associated with the end of ideology movement were embodied in a new class of institutions that came to play a huge role in post-WWII politics—the nonprofit research or advisory corporations—often called "think tanks"—which began as a way to engage civilian researchers in advising on defense-related issues, but which have spread their influence into virtually all policy domains since the 1950s.

The National Academy of Sciences had been created in 1863 in part to advise federal agencies on technical issues, but it was also intended as an honorific society, and prior to the 1960s members never devoted much time to their advising role. The first think tank with a full-time staff exclusively devoted to research and advising on policy issues was the RAND corporation, created in 1948, initially to advise the air force on a variety of tactical and strategic technical issues. By the end of the 1950s there were about a dozen think tanks providing advice to various defense agencies (Smith 1966, 5–6). Today, according to Wikipedia, there are more than 170 think tanks focusing on and significantly influencing federal policy issues related to every governmental interest.

RAND has been the paradigmatic think tank, and by looking briefly at RAND characteristics and activities we can explore some of the ways in which scientistic and technocratic views have shaped American politics and society. The use of technical analyses has been of unquestionably great importance, especially in areas like national security, where technological innovations have made traditional political insights problematic as guides to policy and where demand is virtually inelastic with respect to cost. But there have been accompanying dangers to democracy signaled by President Dwight D. Eisenhower's warning in his farewell speech of January 21, 1961, that an overdependence on such research could lead to a situation in which "public policy could itself become a *captive* of a scientific-technological elite." (Emphasis mine)

Virtually all think tank analyses of policy issues involve the techniques of "systems analysis" which grew out of a marriage of the mathematical techniques of operations research and game theory with emerging practices in systems engineering focused on proposing multiple approaches to design and assessing the relative costs and advantages of each before choosing a preferred design (Simon 1960). Operations research (OR) was developed during WWII to provide a set of techniques to solve well-defined problems of optimizing the operational deployment of known resources. In an important sense OR was just Taylorism enhanced by new mathematical techniques, especially those connected with statistical methods. Thus J. W. Pocock wrote in 1954,

> Taylor and his colleagues were working at comparatively elementary levels where, since relatively little had been done, much could be done with relatively simple techniques of quantitative analysis. Operations research now brings . . . more powerful techniques of quantitative analysis and is an extension of Taylor's basic philosophy to possibilities probably beyond Taylor's vision, far reaching as it was. (Fortun and Scweber, 1993, 624)

Moreover, Operation Vittles, the 1948, system for airlifting supplies to west Berlin, explicitly involved an attempt to provide computable algorithms to determine the best implementation of Taylorite time and motion studies to maximize the tonnage delivered, given interdependent resource allocations (Erickson et al. 2013, 52–70). Historians have also pointed out that the mathematician Merrill Flood, who was among the developers of OR and Game Theory, applied the strategies developed by the Scientific Management expert, Lillian Gilbreth, and described in *Cheaper by the Dozen*, in managing his children (Erickson et al. 2013, 27–28 , 193n1.), so familiarity with Scientific Management was a part of his background.

Among the earliest successful applications of OR were the determination of where to site the small number of operational radars in WWII Britain, given the probability of locations from which German aircraft could come and the probability that certain locations in Britain would be targets, and the determination of the pattern of depths that should be used in dropping a limited number of depth-charges to "kill" a submarine. By bringing the mathematical analysis of complex operations into military decision making, OR played a major role in the Allied victory over Germany, which tended to make its decisions based on tradition and intuition, neither of which was very useful in the face of new technological realities.

Systems analysis uses the techniques of operations research to assess the costs and advantages of specific policy paths when a variety of paths are available, because uncertainties create the need for greater flexibility among choices, when large numbers of interdependent variables are involved, and when objectives are often not well defined. The goal of such analyses is to be as precise as possible in evaluating the costs and consequences of a large number of policies each of which represents a slightly different set of competing assumptions and values, just as it is the aim of systems engineering studies to evaluate the costs and consequences of specific design choices when an uncertain mix of considerations including environmental impacts, short-term vs. long-term costs, redundancy of systems, and so forth, are potentially relevant. For even more open-ended issues in which the various possible responses of "enemies" or other relevant actors may be important, game theory or interactive decision theory—the study of mathematical models of conflict and cooperation between intelligent rational decision makers—came to be utilized, and in many cases, mathematical simulations of conflict and cooperation were even run.

Though operations research, systems analysis, and game theory all had origins before or during WWII, each technique was applied and extended, especially by Herbert Simon, during the period between 1947 and the late 1970s to form the core of the new form of scientific management promoted at most think tanks. As a graduate student in political science at the University of Chicago in the late 1930s, Simon had worked as a research student for

Henry Schultz, who was interested in the application of Scientific Management to municipal governments. Together with Clarence E. Ridley, he published *Measuring Municipal Activities: A Survey of Suggested Criteria and Reporting Forms for Appraising Administration* in 1938, and for several years he continued to focus on municipal administration. Then in 1947 Simon published his most famous work, which generalized and expanded his municipal administration work to cover the characteristics of "rational" decision making in any organization.

Administrative Behavior: A Study of Decision-Making Processes in Administrative Organizations laid out a three-step idealized process—1. Identify and list all alternatives, 2. Determine all consequences resulting from each alternative, and 3. Compare the desirability and efficiency of each set of consequences. In the words of Ronald McKean, an early RAND systems analyst, one must take into account "all possible alternatives and all possible allocations of resources among those alternatives . . . [weighing] the possible impacts of all possible alternatives and the impacts. . .of all events not under the optimizer's control and expectations" (1953, 6).

In practice, of course, one can never be certain that all possible decisions have been evaluated. Nor can one be sure that all consequences have been recognized. And in the final step one has to recognize that what one counts as desirable consequences depends on some explicit or implicit set of values which are used to establish the criteria for desirability. Because the potentially infinite number of alternative decisions and consequences implied in McKean's ideal case is completely unmanageable and even very large numbers of alternatives produced sets of equations that were beyond the calculating capacity of available computers to solve, in the early 1950s Simon developed the concept of "bounded rationality" which explored the ways in which the equations used in an optimizing system could be simplified to make computation possible, including how one could limit the range of variables, decisions and consequences considered and how one could identify the values to use in determining the desirability of various consequences so that they could create a finite system containing a limited number of variables in order to make choices which are "satisficing" if they cannot be optimizing.

Simon's notion of bounded rationality was applicable only to quantitative considerations, but the notion is easily extended to incorporate the simplification of mathematical models by excluding whole categories of considerations. When, for example, systems analysis was used to determine what military bases should be closed in the 1970s, it was decided to exclude all considerations of how legislators would be impacted in terms of reelection.

The specific characteristics of any decision-making body or any client seeking advice will determine how to establish appropriate bounds on rational decision making, and given the overwhelming presence of engineers, economists, mathematicians, physicists, cost analysts, and computer programmers on the staff, for example, early RAND studies applied to military issues focused on "objective" considerations and easily quantified variables, giving little or no weight to historical precedent, instinct, emotion, or the unconscious in positing possible enemy decisions (Ghamari-Tabrizi 2005, 146). One of the frequently employed techniques, for example, was the Delphi method deigned by a group of RAND mathematicians including Olaf Helmer. In this method an initial scenario was established—that is, the U.S.S.R. attacked the United States with 20-kiloton atomic weapons—seven independently operating experts were then asked to estimate how many such bombs would be needed to diminish the munitions output of the U.S. war economy by a given amount by responding to several iterations of a quantitative questionnaire. Qualitative comments were allowed only when a respondent's response to a quantitative question deviated markedly from the consensus value (Tolon 2012, 480). Since its partial declassification in the early 1960s the Delphi method has been extensively used by RAND and other think tanks, including the Stanford Research Institute, to study a wide array of topics including counterterrorism issues, projections of racial tension, and values in K-12 educational institutions (Tolon 2012, 50).

Since its early days RAND has added more social scientists, including historians, sociologists, and psychologists to its staff. During the 1950s, 1960s, and early 1970s, RAND psychologists, including Charles Osgood and Irving Janis, sought to incorporate mechanisms which introduced an understanding of processes that involved irrationality into policy decision making (Erickson et al. 2013, 88–102), but their work had little impact on the mathematical rationality underlying OR, game theory, and systems analysis in general. Economics remained the dominant social science at RAND, and when other social scientists were hired, they tended to be those with a quantitative and "scientific" orientation rather than more humanistic ones for reasons explored by Jefferson Pooley and Mark Solovey (2010, 203–215). In fact, as Solovey has recently argued, the funding available for national security–related social science research pushed the social sciences in an increasingly scientistic and technocratic direction which humanistic and left-leaning critics have characterized as "intellectually impoverished, ethically questionable, and anti-democratic" (Solovey 2013, 58).

The Technocratization of the U.S. Military Leadership

An interesting and important problem arises when the characteristics of advisory organizations differ significantly from those of their clients, leading to mismatches between the bounded rationality explored by the advisory organizations and the culture of its clients. Since the characteristics of the advisory organizations we are discussing inevitably incorporated assumptions which we have labeled scientistic or technocratic, any bounded rationality which they incorporated reflected those assumptions and any decisions that they recommended reflected those assumptions. To the extent that clients accepted the advice of these advisory organizations they had to some degree to modify their own cultures to align them with those of their advisors.

Initially, senior air force officers were quick to reject RAND recommendations that challenged long-held military views (Smith 1966, 104–105 n21). Moreover, Generals Curtis LeMay and Thomas White chafed against accepting the recommendations of those, "often junior in age and experience, who come into Government, for a few years at most, from colleges and foundations" (Ghamari-Tabrizi 2005, 47). But as Bruce Smith has argued in connection with RAND's early military clients,

> The military establishment, traditionally conservative in moving toward the adoption of new weapons and methods . . . recognized that they needed the assistance of civilian scientists, technologists, and analysts to operate a modern defense establishment—and they were forced to obtain this assistance to a certain extent on the scientists' terms. One consequence was that the military services began to adapt their internal procedures to provide an environment conducive to creative innovative activity . . . the military seems to have been converted into a giant engineering establishment. (Smith 1966, 21–22)

Few civilians have shown much concern about the scientization or technocratization of the American military establishment, but the circumstances have produced important negative consequences as one moves up the chain of command toward officers whose strategic decisions must incorporate many non-technical considerations, including some cultural awareness of both our allies and our enemies. While engineering has always played a significant role in the education of officers at West Point and the Naval Academy, curricula at both institutions until recently had always included substantial emphases on politics, international relations, military history, languages, and so forth, and some officers majored in non-technical disciplines. But as Mark Hagerott has argued, largely at the insistance of Hyman Rickover, who initiated the nuclear navy in the 1960s, " the Navy abandoned the 'generalist' model. . . Naval academy admissions criteria and curricula were changed such that specialized technical majors replaced general degrees and universal language education. The restructured processes encouraged officers to value specialized technical expertise over general knowledge, that is, [over] integrated operational, strategic, and cultural knowledge" (Hagerott 2008, abstract). As a consequence, through WWII, American general officers were usually effective strategic thinkers as well as tacticians (Ricks 2012, 40–58), but since WWII the situation has changed.

The Air Force Academy, which was a creature of the Cold War has been even more engineering-oriented than the Naval Academy, and AFROTC students can only major in engineering. Virtually all general officers and admirals who entered the services since WWII come from the service academies. Since the advent of the nuclear navy, virtually every officer who commands a ship must have an engineering degree, and no one who has not commanded a ship can reach the level of admiral. As a consequence, today all air force generals, all admirals, and most army generals—that is, almost every senior officer charged with considering broad strategic decisions—while technically competent, has had very little if any training in cultural awareness, historical awareness, prior knowledge of international relations, or language training. They tend to be overly concerned with tactical as opposed to strategic concerns and overly reliant on advanced technologies. They are relatively ignorant of the cultural settings in which they must operate, nor do they tend to seek advice from career diplomats who may be more knowledgeable about local conditions (Ricks 2012).

The one major exception to this final generalization was a group of army officers who were products of the Social Sciences Department at West Point, among whom David Petraeus was the acknowledged leader. Recognizing that the end game of "nation building" had become a critical military function, this group developed counterinsurgency strategies for small and complex operations in which massed firepower was unuseable and technical superiority was of relatively little value (Ricks 2012, 432–46). Petraeus established a Joint Strategic Assessment Team composed of military officers, embassy officials, and academics in Iraq to both advise him on counterinsurgency strategies and to generate support for those strategies among subordinate officers, among diplomats working in Iraq, and among politicians back home (Kaplan 2013, 260–61).

Petraeus's personal problems have no doubt severely limited the influence of this group, and the specific features some of their counterinsurgency strategies have been of limited value in Iraq and Afghanistan; but more insights from local history and anthropology might make it possible to do better. To expand their breadth of knowledge, Thomas Ricks has suggested that rising officers should be "sent to pursue advanced degrees at elite civilian institutions, where many of their basic assumptions will be challenged," or more radically, "send them to live overseas in third-world countries for a 'sabbatical' year of broadening" (Ricks 2012, 458–59).

Technocracy and the Balance between Legislative and Executive Power

When we turn away from strictly military issues to those national security issues which involve civilian policy makers the situation is quite different. There is little indication that the people involved in U. S. legislative politics are becoming more scientistic or technocratic. If we consider the makeup of the House of Representatives of the 112th U.S. Congress elected in 2011, for example, there were no members among the 435 with practical engineering backgrounds (though Daniel Lipinski has a BS in engineering) and only one member, William Foster, from Illinois, with a background in physics, though there were a handful of MDs.

Secondarily, RAND and most organizations like it serve primarily military and other executive branch clients, carefully avoiding legislative branch patronage. As Bruce Smith writes,

> The possibility of "end runs" around an executive sponsor to Congress could make for intricate and troublesome problems in the sponsor-advisor relationship. RAND is acutely, perhaps overly, conscious of the dangers of this tactic; it has strived to avoid any congressional contacts either at the staff or management level that would lessen sponsor-confidence in the organization. (Smith 1966, 100)

One problem created by this situation is that members of Congress have little or no personal capacity to evaluate the products of think tank advisors. Some members of Congress do hire staff with scientific training, and the Legislative Reference Service of the Library of Congress provides brief analyses of some policy issues with technical components; but the legislative branch of the U.S. government currently has almost no independent in-house way to evaluate the advice produced by think tanks because the Legislative Reference Service has to depend largely on think tank reports.

Between 1972 and 1995 the Office of Technology Assessment did provide Congressional members and committees with independent analyses of complex scientific and technical issues, including acid rain, health care, and global climate change, but it was de-funded during the Newt Gingrich era on the grounds that it duplicated work done elsewhere. There is widespread agreement that the OTA did excellent work, but it worked too slowly to be a valuable resource for congressional decisions, which must generally be done within a two-year cycle (Bimber 1996).

The situation has been vastly different when we consider executive branch policy leadership. In general, under Eisenhower and during the Democratic administrations of Kennedy and Johnson, while elected officials and cabinet level officials except in the Department of Defense, rarely came from scientific or engineering backgrounds, there was a great dependence on policy advice from think tanks and from internal advisory structures composed largely of scientists and engineers. One consequence was the strengthening of the executive branch policy-making role relative to that of Congress. But there has been a second and more disturbing impact on American politics.

In principle, when topics involve policy makers, the views of any particular group—supposedly including those presented by a think tank—may be opposed by any interest group with antagonistic views. Thus Smith argued that,

> The pluralism of the advisory system, in which RAND is only one institution among many with access to persons of authority, helps assure that no one group will monopolize the attention of policy makers. . . . The presence of a number of advisory institutions like RAND helps to assure decision makers of a broad base of scientific advice and to guard against the dangers of a closed system with a narrow technocracy cut off from the healthy effect of outside criticism. (Smith 1966, 314–15, 321)

But executive branch clients have shown little or no willingness to fund multiple competing analyses of issues, and in practice there is a huge asymmetry between the access which think tanks have to policy makers and the access which potential critics might have. For example, in 1959, when the RAND researcher, Herman Kahn, had developed the analyses of possible thermonuclear war scenarios which appeared the next year as *On Thermonuclear War,* he went on tour, giving three-day presentations to over 5000 policy makers and advisers (Ghamari-Tabrizi 2005, 206–07). Persons with countervailing opinions were lucky to get twenty minutes with a handful of congressional staffers. Similarly, in connection with its most famous 1953 study, which challenged the initial air force request for an analysis of how to maximize the security of B-52s in advanced bases so as to maintain a credible second strike threat to the Soviets, RAND's Albert Wohlstetter had no fewer than ninety-two briefings with the air force within a seven-month period during which he promoted the development and deployment of hardened ICBMs while no independent outside views were sought (Green 1968, 316).

It is certainly true, especially in cases not directly related to issues connected to national security, that political considerations have often led to think tank recommendations being ignored or to decisions not to have think tank analyses in the first place. On almost all issues connected with national security, however, and at least some non-security issues, Eisenhower's warning that "public policy could itself become a *captive* of a scientific-technological elite (Eisenhower 1961)" was, at least during the 1960s and early 1970s, a serious concern among both academics and politicians (Wang 1999, 282–287).

Scientism and Technocracy in the Kennedy Administration

The administration of John F. Kennedy was in many ways the most scientistic and technocratic in U.S. history. In part because of his own preferences and in part because when he ran against Richard Nixon he could not afford to be or seem soft on Communism, John Kennedy distanced himself from traditional liberal Democrats like Adlai Stevenson, Hubert Humphrey, and Chester Bowles. At the same time, he could not seem too close to traditional Republican anti-communists or he would have lost what liberal support he gleaned for his domestic policies. His response was to surround himself with bright young men with little party political experience. For the most part they had all been involved with the military during WWII and agreed on one feature of foreign policy—that totalitarianism had to be stopped and that it was necessary to use force to do so. But according to David Halberstam, the thing that bound them together even more closely was a kind of arrogance and "the belief that sheer intelligence and rationality could answer and solve anything" (1972, 44).

Perhaps most influential within the group was McGeorge Bundy, a Yale graduate in mathematics from a wealthy New England Republican family background, who had spent the war in the navy as an Admiral's Aide. He had served on the Council on Foreign Relations study group on the Marshall Plan which directed covert CIA activity in Europe beginning in 1949 and then became Professor of Government at Harvard. In 1953 he was appointed Dean of the Faculty of Arts and Sciences at age thirty-four with nothing more than his bachelor's degree and a reputation as a brilliant intellect and administrator. Kennedy, who had graduated from Harvard, offered Bundy positions in the State Department and Department of Defense, which he declined, before Kennedy made him Special Assistant to the President for National Security Affairs, a position which gave him virtually unlimited access to the president and immense power. Combining a brilliant mathematical mind with knowledge of covert operations and international relations he was easily capable of digesting the results of operations research and systems analysis and using them to guide policy. According to the political writer Joseph Kraft, writing in 1965 before the Viet Nam War had clearly veered out of control,

> Bundy is the leading candidate, perhaps the only candidate for the statesman's mantle to emerge in the generation that is coming to power . . . His capacity to read the riddle of multiple confusions, to consider a wide variety of possibilities, to develop lines of action, to articulate and execute public purposes, to impart quickened energies to men of the highest ability seems to me unmatched. (Halberstam 1972, 46)

Bundy brought Walter W. Rostow into the White House as his deputy. The son of Russian Jewish immigrants who identified themselves as Socialists, Rostow was a scholarship student at Yale, where he got his bachelors, masters, and PhD degrees in Economics. During the war he served in the Office of Stategic Services using OR to choose targets for allied bombing runs. Then after short stints teaching at Oxford and Cambridge and serving as an economist for the State Department, in 1950 he became Professor of

Economic History at MIT and a vehement advocate of free-market economics. Kennedy had known Rostow and used him to provide position papers on a number of issues during the late 1950s, and he admired his intellect. As we will see later when we consider his model of development, fully expounded in *The Stages of Economic Growth: A Non-Communist Manifesto* (1960), his interests were very much theoretical as well as historical.

Only slightly less influential with Kennedy than Bundy, and even more technocratic, was the man chosen by Kennedy to head up the Department of Defense. Robert McNamara was not personally known to Kennedy before several advisors recommended his appointment, but he soon came to have close personal relations with the entire Kennedy family. McNamara had been born in San Francisco, California, in 1916, son of the sales manager for a wholesale shoe manufacturer. In 1937 he graduated from U.C. Berkeley in economics with minors in mathematics and philosophy—which at the time meant logic and analytic philosophy. Two years later he got an MBA degree from Harvard. After a year at Price-Waterhouse, he was back at Harvard, teaching courses on applying analytic techniques of business management to both regular students and to air force officers in a special program.

During the war McNamara served in the air force's Office of Statistical Control, analyzing the efficiency and effectiveness of bombers in the Asian theater. At the end of the war, he and ten colleagues from the Office of Statistical Control were hired by Henry Ford II to revamp the struggling Ford Motor Company. He started in 1946 as manager of planning and financial analysis, and McNamara and his "whizz kid" friends used their analytical skills and some shrewd market studies that showed an important place for smaller cars in the American and European markets to turn money-bleeding Ford into a profitable and well-managed company. By 1960 he had taken on a series of management positions at Ford, and that October he became the first non-family member to become president of Ford Motor Company. Just a few weeks later he was tapped to become Kennedy's head of the Department of Defense.

Inside the administration McNamara was known as "the computer" because of his dispassionate, analytical approach to all policy issues, and his testimony before Congress sometimes sounded as if it could have been produced by a machine. Unsuccessfully defending his use of computer calculations to delay development of the B-70 bomber against critics who condemned him for accepting, "the conclusions of nonmilitary intellectuals who take the word of inanimate computers over the experience, training, and judgment of military professionals," McNamara simply laid out the systems-analysis method of coming to his decision:

> In contrast to most other military requirements, the requirement for strategic retaliation forces lends itself rather well to reasonably precise calculations. . . .

> The first step in such a calculation is to determine the number, types, and locations of the aiming points of the target system.
>
> The second step is to determine the numbers and explosive yields of weapons which must be delivered on the aiming points to ensure the destruction or substantial destruction of the target system.
>
> The third step involves the determination of the size and character of the forces best suited to deliver these weapons, taking into account such factors as:
>
> 1. The number and weight of warheads that each type of vehicle can deliver.
> 2. The ability of each type of vehicle to penetrate enemy defenses.
> 3. The degree of accuracy that can be expected of each system.
> 4. The degree of reliability of each system, i.e., the proportion that we can count on getting off successfully within the prescribed time.
> 5. The cost-effectiveness of each system, i.e., the combat effectiveness per dollar of outlay. . . .
>
> With these considerations in mind, and utilizing the analytic procedures I outlined earlier, we studied a large number of alternative combinations of weapons systems and finally arrived at the force structure proposed for the next few years. This process led to the conclusion that the proposed B-70 would not justify its high cost (*New York Times*, March 28, 1962).

Had McNamara emphasized his own war-time experience in the air force managing the project that developed the B-29 then evaluating tactics and weapons systems using systems analysis, he almost certainly would have been more effective in responding to his critics than he was in giving congressmen a primer course in its techniques.

Even before he took office, McNamara had recruited Charles J. Hitch, who had left a position as the first American Rhodes Scholar to be invited on to the Oxford University faculty, in order to create the Economics Division of RAND in 1948, to become Assistant Secretary of Defense and Comptroller of the Department of Defense. At RAND, Hitch had specialized in defense economics and had written, with Roland McKean, *The Economics of Defense in a Nuclear Age* earlier in 1960. Moreover he had been elected president of the Operations Research Society of America in 1959. McNamara brought him into the Defense Department to install his systems-analysis-based budgeting procedure, the Planning, Programming, and Budgeting System (PPBS) to all defense-related expenditures (Novick 1962). In spite of determined opposition by the services and by military contractors the new system was successful in increasing the efficiency of Pentagon operations and reducing waste. One estimate suggested that it saved $4.1 billion in its first year of implementation (Nieburg 1966, 356).

When Hitch's successor at both RAND and the DOD, Alain Enthoven, wrote a tribute to Hitch after the latter's death; he characterized Hitch in a way that intentionally expressed the very real strengths and unintentionally exposed the weaknesses of all of the technocrats among Kennedy's closest advisors:

> He had a backbone of steel. He considered his positions very carefully, then held them very firmly. He stood up to enormous pressures. At RAND and the Defense Department, Charlie found it appropriate to recruit self-confident (some said arrogant) young men who liked to ask hard fundamental questions, gore sacred cows, attack the conventional wisdom and the vested interests. [We] had little sympathy for a congressman who didn't want an air base or a factory in his district closed. This led to frequent demands for their scalps. Charlie protected them, stood up to the powers that be, insisted that the hard questions be asked and that the answers be based on the evidence and the merits. (Enthoven 1995)

There was a kind of admirable integrity about all members of the group and an often valuable willingness to challenge some traditional assumptions (except the assumption that there existed a monolithic Communist movement devoted to world dominance) along with an unwillingness to change their minds, once made up that was less admirable. Moreover, their decisions were virtually always based on evidence. But what they counted as evidence was problematic, because they virtually refused to count as evidence anything that could not be quantified, like the political realities facing congressmen or human factors that the longtime experience of career military men or diplomats understood.

Given accurate information and their analytic abilities Kennedy's technocrats made some very important positive policy decisions. For example, when McNamara discovered that there was no real missile gap between the Soviet Union and the United States, he sought to reign in the race to produce ever more missiles. Even more importantly, the Eisenhower administration had promoted a strategy of massive retaliation which would initiate the use of tactical nuclear weapons to make up for a planned reduction in manpower. But when McNamara's teams of analysts ran war games, no matter which side used nuclear weapons first, retaliation always created a scenario of escalation to the use of strategic weapons with massive casualties among civilians on both sides. This result was NOT rational, so McNamara carefully began to insist that neither we nor our allies use tactical nuclear weapons—satisfying hawks by building up conventional troop numbers. Similarly, he fought to slow down the production of nuclear missiles and worked tirelessly for the first nuclear test ban treaty.

The focus on quantifiable evidence was, however, often dangerous in part because, as was true in the Soviet Union, those who provided the evidence either unintentionally or intentionally provided numbers that did not accurately reflect the situation. Thus, for example, during the Viet Nam conflict, American military commanders and embassy officials padded estimates of

Viet Cong casualties as well as the damage to enemy supply routes done by Air Force bombing raids to advance their own careers. This falsification of evidence created the impression that U.S. strategy was vastly more successful than it really was and encouraged McNamara to continue to promote a military strategy long beyond the point at which many career military officers had insisted on its ineffectiveness.

The other problem with an over-reliance on quantification, which assumed a one size fits all approach to tactics and strategy, is that it encouraged Kennedy's policy makers to either not seek or to ignore context-dependent and often non-quantifiable knowledge which those who had prior experience either could have offered or did offer. David Halberstam reports a conversation that Kennedy had with his Secretary of Labor, Arthur Goldberg, after the tragic CIA-sponsored Bay of Pigs invasion of Cuba, in which the United States promoted and supported an invasion of Cuba via an amphibious landing by an anti-Castro Cuban guerrilla force of approximately 1,400 persons. In this case the issue was a tactical one. Goldberg approached Kennedy to complain that the president and his close advisors led by Bundy, McNamara, and Robert Kennedy should have consulted more widely. After Kennedy chided him because his post was in Labor, not Foreign Policy or Defense, Goldberg replied:

> You are making the mistake of compartmentalizing your cabinet. There are two people in the Cabinet you should have consulted on this one, men who know some things, and are loyal to you and your interests.
>
> Who? Kennedy asked.
>
> Orville Freeman [Secretary of Agriculture] and me.
>
> Why Orville?
>
> Because he's been a Marine, because he's made amphibious landings and because he knows how tough they can be even under the best of circumstances. He could have helped you.
>
> And why you?
>
> Because I was in OSS during the war and I ran guerrilla operations and I know something about guerrillas. That they're terrific at certain things. Sabotage and intelligence, nothing like them at that. But they're no good at all in confronting regular units. Whenever we used them like that, we'd always lose all our people. They can do small things well, but it's a very delicate, limited thing. But you didn't think of that—and you put me in the category of just a Secretary of Labor. (Halberstam 1972, 71)

While Goldberg may have made these comments out of personal pique, the point that he made about the narrowness of consultation on the part of Kennedy's closest advisors is correct.

While the State Department and Joint Chiefs of Staff were nominally involved in the decision to go ahead with the Bay of Pigs plan, strenuous objections from both sources based on tactical, strategic, and moral grounds were simply ignored. Old Cuba hands insisted that the vast majority of Cubans supported Castro and that an invasion supported by the United States would simply cement that support; others argued that the U.S. provision of supplies and training for the invasion of Cuba was a violation of any reasonable moral code; still others pointed out that the proposed invasion force—especially since Kennedy was unwilling to chance the use of American Air Force support lest it precipitate a much wider conflict—was far too small to be successful (Halberstam 1972, 66–71).

Even more egregious was the lack of local knowledge of conditions in Southeast Asia which led to decisions about the war in Viet Nam. There were several well-placed State Department officials including Chester Bowles, Under Secretary of State, and Kenneth Todd Young, who had served in Viet Nam in the past, as well as George Ball, soon to replace Bowles, who opposed continuing support of the Diem regime because they knew well that Ngo Dinh Diem and his brother Ngo Dinh Nuh were corrupt and widely disliked by the Vietnamese people. On the contrary, the communist Vietminh had led in the fight to oust the French and were viewed as heros by most Vietnamese peasants. But Young was passed over for the ambassadorship to Viet Nam in March of 1961 in favor of Frederick Nolting, who had never spent time in Asia and who was a well-known anti-communist who saw Diem as the only viable Vietnamese anti-communist leader. In addition, Nolting not only did little to encourage his staff to look for information beyond that fed to him by the Diem government, he actually forbade embassy staff members to talk with Vietnamese dissidents (Halberstam 1972, 132).

Vice President Lyndon Johnson, who also had no Asian experience was sent on a public relations trip to the Far East, and though he did not like Diem personally, he came back from the trip claiming that communist expansion had to be stopped and that there was no alternative to Diem to lead the effort. Finally, in order to protect his right political flank, Kennedy had appointed the vehement anti-communist John McCone as director of the CIA, which did little to balance the pro-Diem views of Nolting and Johnson.

Under pressure to expand the advisory presence of the United States in Viet Nam, in late 1961 Kennedy decided to send his own special representatives to Viet Nam on a fact-finding mission. That mission was led by Walt Rostow, whose WWII experience picking bombing targets gave him an assurance, never touched by contrary evidence, that bombing could be an effective weapon in Viet Nam. By 1961 Rostow had also become a vehement anti-communist, so he was inclined to support the Diem regime. The other major figure in the mission was General Maxwell Taylor, Special Military Representative of the president. Taylor was well read, intelligent, and articulate, and he had recognized long before many military men that after the Korean Conflict most wars would be what he called "brush fire wars," which Kennedy understood to be guerrilla wars.

Though Taylor had opposed Eisenhower's original commitment of 200 advisors to Viet Nam, he had also opposed the Eisenhower broader commitment to massive retaliation via atomic and nuclear weapons. Furthermore he had served in an airborne capacity in the army and shared Rostow's confidence that air power could be exceedingly successful in Viet Nam. He and Rostow agreed that Hanoi would be vulnerable to bombing, but he was also optimistic about the role of airborne troops in guerrilla warfare, because it would make conventional troops with their superior technology highly mobile. Neither he nor Rostow thought that the geography of Viet Nam, the support which communist troops would have among the peasants, and the willingness of the Viet Cong to accept massive casualties, could overcome technical superiority.

Though it was not widely known at the time, the Rostow-Taylor report recommended the immediate commitment of U.S. combat troops to Viet Nam, but Kennedy chose to move more slowly, providing more advisors (15,000) to train South Vietnamese troops and providing more equipment—especially helicopters and light aircraft to increase their mobility. Thus, while Kennedy was more cautious than his leading advisors, he did escalate the American presence and created pressure for ever more involvement. Through 1962 and 1963 a few military commanders in the field were willing to risk their careers to honestly report on the ineptness and corruption of the South Vietnamese military, but their attempts to reach the Joint Chiefs of Staff were blocked by Taylor.

Starting in 1962 Robert McNamara began to take the lead among Kennedy's advisors on issues concerning Viet Nam. Knowing nothing about Viet Nam and its people, NcNamara took several trips to Saigon to find out the truth of what was going on directly. But like Nolting's staffers, he was carefully kept away from dissidents and he accepted the false statistics being generated by Commanding General Paul Harkins and the embassy, plugging them into his rationally grounded analytic models and insisting that the Viet Nam war was being won. Finally, in 1965, still head of the Defense Department under Johnson, McNamara oversaw the biggest escalation of the war. Eventually, however, "when the mathematical version of sanity did not work out, when it turned out that the computer had not fed back the right answers and had underestimated those funny little far-off men in their raggedy pajamas, he would be stricken with a profound sense of failure, and he would be, at least briefly, a shattered man" (Halberstam 1972, 218). In this case the Kennedy technocrats' ability to make a decision and stick with it and their dependence on quantitative data that they did not collect themselves had tragic consequences.

The Special Role of the President's Science Advisory Committee

On a much happier note, the Kennedy administration amplified an Eisenhower-initiated apparatus within the Office of Science and Technology in the White House that was capable of evaluating and sometimes counteracting outside technical advice. Eisenhower had created the position of President's Science Advisor and installed first James Killian, president of MIT, and then Harvard chemist and Manhattan Project veteran, George Kistiakowsky, in that position. Furthermore, he had created the President's Science Advisory Committee (PSAC) with direct responsibility to the president to undertake independent in-house analyses of issues with major technical components. Under Eisenhower, Kistiakowsky and PSAC were consulted almost exclusively on space and defense issues. But under Kennedy, the president's science advisor, Jerome Wiesner, had direct and daily contact with the president and he was able to bring new issues to the president's attention. Moreover, PSAC was much more frequently asked to study a wide variety of issues, to some degree mitigating at least one of the problems of the de-facto contracting out of policy decisions to extra-governmental advisory corporations.

Ironically, more often than not PSAC reports threw doubts on the technology-promoting reports of outside advisors and military proponents, possibly in part because as highly independent part-time advisors whose only compensation was to have their expenses paid, committee and subcommittee members were more likely to question assumptions that were ideologically or politically motivated because they did not fear retribution. Eisenhower even said just before he died that, "this bunch of scientists was one of the few groups that I encountered in Washington who seemed to be there to help the country and not help themselves," and he viewed them as balancing the military-industrial complex which he saw as such a threat (Wang 2008, 2).

As veterans of the Manhattan Project or the development of radar, most members of PSAC were convinced that technological issues could not be divorced from social and political concerns, so they did not even pretend to advise on technical issues alone. They willingly considered a wide range of policy options and advocated for their preferred choices, though on virtually all issues during the Kennedy administration, PSAC members and those that they recruited onto panels to report on specific issues respected the need for confidentiality in their reports to the president. When, however, they felt that Richard Nixon intentionally distorted the advice which they provided opposing investment in developing a supersonic commercial airliner and in promoting the Sentinel antiballistic missile system (ABM), some members broke ranks and went public. Nixon's angry response was to dismantle the entire White House science advisory apparatus (Wang 2008, 287–308).

Though it was nominally resurrected by George H. W. Bush in 1990 as the President's Council of Advisors on Science and Technology (PCAST) it has subsequently had very little of the clout that PSAC had under Eisenhower and Kennedy.

Chapter Five

Technocracy/Scientism in the Developing World, 1900–2000

In order to understand the rise of technocrats in the developing world of the twentieth century it is important to briefly characterize European imperialism during the late nineteenth century. With few exceptions European nations involved themselves in colonial expansion during the nineteenth century primarily to provide markets for their own industrial production and secondarily to stabilize the supply of raw materials for industrial production as well as food supplies for populations that had developed tastes for tropical products—coffee, tea, chocolate, bananas, and so forth. The French politician, Jules Ferry, explained the first part of the situation beautifully in a speech to the National Assembly on July 28, 1873:

> In the area of economics I will allow myself to place before you, with the support of some figures, the considerations which justify a policy of colonial expansion from the point of view of that need, felt more and more strongly by the industrial populations of Europe and particularly those of our own rich and hard working country: the need for export markets . . . Yes, what is lacking for our great industry, drawn irrevocably on the path of exportation by the [free trade] treaties of 1860, what it lacks more and more is export markets. (Andrea & Overfield 1990, 307)

Even when European nations seemed to be offering aid to dependent colonies during the late nineteenth and early twentieth centuries, that aid was virtually always tied to two major goals of the European colonizers. First, it was almost always intended to promote the monoculture of crops—cash cropping—sought by the colonial power, in spite of the fact that cash cropping often had a triply negative consequence for the local populations. It reduced the local economy's ability to feed its own population by taking the best land out of food production, it sometimes led to the rapid impoverishment of the soil, and it almost inevitably produced dramatic income concentration, leaving large local peasant populations worse off than they had been before, while enriching small elites of relatively wealthy natives or European colonists. (Villianotis 2006, 213—229)

In a number of African colonies, for example, British aid was made available in the early twentieth century to create railroads to carry commercial crops to market rather than to improve locally more important internal road networks to connect local and regional food markets. Moreover the aid, largely in terms of insured or reduced interest private loans, was made available on the condition that the borrowing colony purchase a large fraction of the rails and rolling stock, including locomotives, from suppliers back in the European lending nation (Burch 1988, 179). Just one of the major consequences of this pattern was that when the colony eventually became self-ruling, it was saddled with a huge debt service which both crippled its immediate abilities to provide services such as health care and education to its population (Burch 1988, 202 n1)—and which also had the long-term effect of transferring capital from the underdeveloped South to the industrialized North (Peng 1988, 211–212).

Technocrats and Early American Imperialism

Between the end of the Mexican American War, which established the Rio Grande River as the southern boundary of Texas and added the territory which subsequently became New Mexico, Arizona, Nevada, and California to the United States, and the late 1890s, American expansion had not involved military conquest; nor had it extended beyond the continental boundaries of North America. The Oregon Territory south of the 49th parallel had been co-occupied by the United States and Britain since 1818, but American settlement had overwhelmed British interests, so Britain ceded its interests in 1845 for a $30 million dollar payment, and Secretary of State Seward negotiated the purchase of Alaska in 1867 for $7.2 million.

While the expansionist Seward proposed the annexation of the Hawaiian Islands and Cuba and promoted President Grant's proposal to annex the Dominican Republic in 1870, congressional opposition to overseas expansion was sufficiently strong to block these moves. During the next quarter century, however, several circumstances increased both expansionist sentiments and opportunities. First, American industrial production increased rapidly; so by the mid-1890s the United States had the same economic motives that had fueled European imperialism from midcentury; it needed export markets to absorb the products of its factories. Thus, when the question of annexing the Philippines came up in 1898, Senator John Beverage of Mississippi addressed President McKinley in a speech to the 56th Congress:

> Our largest trade henceforth must be with Asia. The Pacific is our ocean. More and more Europe will manufacture the most it needs, secure from its colonies the most it consumes. Where shall we turn for the consumers of our surplus? Geography answers that question. China is our natural customer. She is nearer to us than England, Germany, or Russia, the commercial powers of the present and the future. . . . The Philippines give us a base at the door of all the East. (Beveridge 1898, 704)

Second, the British Social Darwinist, Benjamin Kidd, articulated a new economic and moral doctrine in his *Social Darwinism* of 1894 and *The Control of the Tropics* of 1898. Kidd initially seemed to agree completely with earlier colonial theorists. In order to continue the economic growth of Europe and the United States it would be necessary to expand the exploitation of the natural resources of the tropical regions of the world, which contained roughly two-thirds of the natural resources needed to maintain the industrial growth of the civilized world. Because of their lack of "social efficiency" the native populations of the tropical regions were unable to do the job. According to Kidd, "Over a considerable portion of these regions at present we have existing a state either of anarchy or of primitive savagery, pure and simple in which no attempt is made or can be made to develop the natural resources lying ready to hand" (Kidd 1898, 15). Such a situation could not be allowed to continue: "The last thing our civilization is likely to permanently tolerate is the wasting of the resources of the richest regions of the earth through the lack of the elementary qualities of social efficiency in the races possessing them" (Kidd 1898, 96).

Unfortunately, from Kidd's Christian perspective, all traditional forms of European imperialism were unacceptable because of the way that they brutally exploited native populations and robbed them of their rightful possessions. So some new form of imperialism had to be instituted—one which would simultaneously honor the interests of indigenous peoples and allow for the development of their resources: "The right of those races to remain in possession will be recognized; but it will be no part of the future conditions of such recognition that they shall be allowed to prevent the utilization of the immense natural resources which they have in charge" (Kidd 1998, 96–97).

To achieve these dual goals Kidd argued that civilized nations must govern the tropical regions as a *trust* for the sake of both the civilized world and that of the native populations. The governments instituted by the civilized country would ensure that natural resources were efficiently developed "under the principles of our Western individualism" (Kidd 1898, 57) and that the benefits accruing from the sale of such resources would be fairly shared with the native populations. Furthermore, in Kidd's view, the governing trustee nation would not seek to limit access to the resources in its own national interest, but would open them up to all civilized nations.

Like Darwin, but unlike many social Darwinists, Kidd did not believe that any human race was intrinsically superior or inferior: "Neither in respect of colour, nor of descent, nor even of the possession of high intellectual capacity, can science give us any warrant for speaking of one race as superior to another" (Kidd 1898, 98). Since racial differences were a consequence of social evolution they might eventually be reduced or eradicated through cultural contact. But Kidd did not expect the process to be rapid; so he envisioned the trusteeship process as an extended one.

American expansionists such as Franklin Giddings, Bernard Moses, William McKinley, and Woodrow Wilson took up Kidd's fundamental arguments. They were, however, vastly more optimistic about the abilities of indigenous cultures to adapt rapidly—possibly within a few years. Moreover, they saw the United States as uniquely situated to extend the virtues of civilization to tropical peoples. According to Giddings, only recently in the Western world had "the thought of an ever continuing emancipation and enlightenment of the whole human race" become a common ideal (Giddings 1901, 334). For a variety of reasons other Western nations had not yet implemented that full vision, but now, "The American, who through liberty and present acceptance of the practical point of view, has achieved a worldly success, stands ready in the spirit of fraternity to reach out a helping hand to the brother who has not yet succeeded, and to aid him in every possible way to attain the objects of his desire" (Giddings 1901, 339). The idea of liberty had been developed largely among the English and that of equality had been the gift of the French. It was now the task of America to combine and universalize those ideals (Giddings 1901, 340).

Events associated with Cuban resistance to Spanish rule led to the Spanish American War, which presented Americans with the opportunity to try out the ideas being promoted by expansionists. In 1898, at the end of the war, Spain ceded Puerto Rico, Guam, and the Philippines to the United States and decisions had to be made about what to do with them. President McKinley took the position promoted by Kidd and his followers but modified by Giddings. Speaking in early 1899 McKinley argued, "The Philippines are not ours to exploit, but to civilize, to develop, to educate, to train in the science of self government" (Go 2007, 76). Similarly, Secretary of War Elihu Root, speaking of both Puerto Rico and the Philippines said, "it is our unquestioned duty to give them . . . to the greatest possible extent freedom, self government . . . and development in civilization" (Go 2007, 83). While the Foraker Act of 1900 put ultimate power in the hands of American governors in both the Philippines and Puerto Rico, it called for political education of the native populations and for the involvement of elected local officials in policy discussions.

Both from an early twenty-first-century Western perspective and from the perspective of Filipinos who had been fighting against Spanish rule in 1898, these American ideals appear(ed) to be appallingly paternalistic and insensitive to the values of other cultures, and they led to a continuing Filipino resistance to American governance for several years. But at the time they seemed to both many Americans and Europeans to be extremely progressive, and foolishly so. That American government on the ground actually reflected these ideals to a large extent is suggested by the ascerbic comments of European observers. The wife of a British diplomat in Malaysia, for example, wrote to her diary, that the Americans act as if, "every people in the world should have self-government and equal rights. This means . . . they consider these Malay half-breeds to be capable . . . of understanding the motives and profiting by the institutions which it has taken the white races two or three thousand years to evolve" (Go 2007, 82). Her judgment was that the American "fad" would soon give way to a more sensible approach to colonial governance; but in spite of continued resistance to American government on the part of some Filipinos, by 1907 the Philippines had elected a National assembly with substantial power and local government was largely under local Filipino control.

Most discussions of American exceptionalism with respect to imperialism have focused on the export of political democracy—a view which was still being expressed by George W. Bush in 2003 to justify American intervention in Iraq. But recently, James C Scott (1998) and Michael Adas (2006) among others have turned the focus toward the implications of the other side of Kidd's argument—that is, that Western government in the tropics had to be aimed at the efficient exploitation of tropical natural resources for the purpose of promoting the economic development of the "civilized" world as well as for the benefit of the native populations. With little or no regard for local cultural values, Americans evaluated native uses of resources only against American standards of resource exploitation, with their dependence on intense work patterns, industrialization, and advanced technologies.

American expansionists saw engineering experience as an excellent background for Colonial administration. According to "Publicola," who's *Duty of the American People to the Philippines* (1898) reflected both the McKinley administration's views and those of the subsequent Roosevelt administration,

> Our railroads and great iron establishments will be found not bad schools from which to graduate administrators for provinces and districts. There are hundreds of young engineers with some practice in the field who would go out for minor positions with the same sense of duty and very nearly the same education as that taken to India by Lawrence, Edwards and Roberts. (Adas 2006, 145)

American administrators did call for the scientific management of resources, they emphasized the material betterment of the Philippine people even over their political participation, and they engaged large numbers of engineers, with their technocratic attitudes, in colonial governance. In fact, when civilian government was instituted in the Philippines in 1901, although the overall governor was William Howard Taft, a political appointment with governmental experience, it was mandated that the supervisors of the twenty-seven Philippine provinces be civil engineers (Adas 2006, 145). Young engineers played a central role in establishing the infrastructure of public works—especially of the roads and sources of power that were prerequisites for resource development. W. Cameron Forbes, the fourth governor of the Philippines, who served from 1909 to 1913, acknowledged the role of American engineers when he reflected on the early days of American rule:

> In the main it may be said that of the many creditable and useful services rendered by Americans in the Philippine Islands that given by the Bureau of Public Works has been among the foremost. The Engineers, from those in the highest positions down through the service, brought to their work true American enterprise. . . . Too high praise cannot be given to the devotion revealed and service rendered by the band of pioneers who, as builders and planners, did their part in the fine work of civilization the United States undertook to carry out in the Philippine Islands. (Adas 2006, 146)

Furthermore, education beyond basic literacy and citizenship in the newly acquired lands was oriented primarily toward technical training, science, and engineering.

Under Spanish rule technical education of any kind had been discouraged. In fact, the Dominican rector of the single University in the Philippines during the nineteenth century held that, "Medicine and the natural sciences are materialistic and impious studies" (Alcala 1999, 117). Under American rule, vocational schools patterned on Booker T. Washington's early Tuskegee Institute were established (May 1980, 89–92), though they were not well attended (Caoili 1986, 23). While the first institution of higher education established by the Americans in 1901 was the Philippine Norman School which trained teachers, a Medical School was established in 1905, and in 1908 the University of the Philippines, which soon offered programs in agriculture, veterinary medicine, engineering, public health, forestry, and pharmacy was created (Alcala 1999, 118–119).

In the meantime, in order to develop Filipino professionals and potential faculty for the university, starting in 1903 the American civil government provided scholarships for 135 Filipinos to study engineering, medicine, and education at schools in the United States (Caoili 1986, 23). This program was immensely successful. Between 1911 and 1925, for example, the tenure track Filipino faculty at the university increased from 14 percent to 78 percent

(Caoili 1986, 23 n 67). Between 1901 and 1913, Filipino engineers in the Bureau of Public Works went from zero to 18 out of 145 and by 1925 they numbered 174 out of 190 (Caoili 1986, 25).

While American political leaders including Theodore Roosevelt argued that civilizing the Philippines should involve encouraging industrial development, both local Philippine elites and American capitalists were vastly more interested in promoting more efficient agriculture, timber production and mineral extraction—that is, in maintaining the Philippines as a supplier of raw materials and as a market for American manufactured goods. Consistent with Scientific Management in the United States, the emphasis in the Philippines was in increasing the production of commercial crops, timber, and minerals without consideration of issues of social justice or long-term sustainability. Furthermore, consistent with political applications of Scientific Management in Europe, centralized planning, rather than local control was seen as a viable strategy for maximizing overall efficiency (Gauld 2000, 232).

Development and the Cold War: Truman, Rostow, and the Soviets

In his inaugural address of 1949, Harry S. Truman gave one of his most important speeches, explaining that the United States was about to take on a new role in assisting the economic development of countries needing aid:

> More than half the people of the world are living in conditions approaching misery. Their food is inadequate, they are victims of disease. Their economic life is primitive and stagnant. Their poverty is a handicap and threat both to them and to more prosperous areas. For the first time in history humanity possesses the knowledge and the skill to relieve the suffering of these people . . . I believe that we should make available to peace-loving peoples the benefits of our store of knowledge in order to help them realize their aspirations for a better life . . . Greater production is the key to prosperity and peace. And the key to greater production is a wider and more vigorous application of modern scientific and technical knowledge. (Truman 1949)

While Truman acknowledged that economic development was ultimately about the well-being of people—including their health, access to education, social inclusion, safety, and so forth—and not just about economic growth, the final sentence above indicates that he simply assumed that economic growth—that is, increased aggregate productivity—was the foundation upon which other features of development must be grounded. Furthermore, he assumed that the application of advanced science and technology would be the basis of economic growth (Rosenberg 1970, 550).

On the face of it, aid to developing countries would now be decoupled from the interests of those offering help, including the kind of spending requirements that had limited the value of aid in the past. But aid for development was more than simply altruistic. U.S. and Western European aid would now be used as a tool in the Cold War to bind receiving countries to their benefactors. By the same token, The U.S.S.R., which now saw technology as a key tool in the growth of Soviet-style Communism, sought to use its aid for much the same purpose. For present purposes we will ignore military aid and focus exclusively on aid at least nominally aimed at economic development projects.

Through the Truman and Kennedy administrations, "development" was a relatively loosely understood term with a general assumption that each nation might find its own path toward its own goals. Both the U.S. AID program, which provided engineering expertise, and the Kennedy administration's Peace Corps program allowed developing nations to propose their own projects with relatively little guidance, though economic growth was a significant consideration. During the early 1950s, for example, the government of South Korea utilized most of its American aid to increase domestic consumption rather than to directly increase production (Scitovsky 1985, 222). Within the Kennedy administration the Cold War emphasis was on promoting constitutional democracy through both social and economic reform and on opposing perceived tyrannies of both the right and the left. Thus, as late as 1962, when a right wing military coup overthrew the democratic government of Peru, the United States cut off all military and economic aid (Muller 1985, 448).

Under Johnson, anti-communism became vastly more important than the support of democracy. Thus Johnson's Assistant Secretary of Sate for Inter-American Affairs and Special Assistant to the President announced the henceforth, Kennedy's Alliance for Progress would focus, not on supporting democracies, but on "economic growth, protection of American investments, non-intervention, and opposition to Communism" (Muller 1985, 449). The anti-Communist focus remained during the Nixon and Reagan administrations whose development policies were guided by the views of Walter W. Rostow as expressed in Rostow's *The Stages of Economic Growth: A Non-Communist Manifesto,* published in 1960 and revised in 1971 and 1990.

Rostow's text purported to be an historical analysis of the path by which virtually all Western industrialized nations had reached their status as well as a blueprint for the path by which presently underdeveloped nations could expect to achieve their appropriate place in the capitalist "free market" system—the presumed natural goal of development (thus the subtitle of Rostow's work). For Rostow and his followers, economic growth came to constitute and not simply underlie economic development, and virtually all aid proposals were evaluated in terms of their potential to promote economic growth. So, for example, between 1965 and 1981, aid to Korea was refocused almost exclusively on increasing productivity (Scitovsky 1985, 222).

Most development economists have a much broader understanding of economic development, which focuses on what has come to be called "human development," a concept that incorporates all aspects of individual well-being, and they clearly distinguish that concept from mere economic growth as Truman did implicitly. In 2011 the United Nations Development Programme initiated a Human Development index which includes measures of life expectancy at birth, mean years of schooling, and gross national income per capita, but individual development economists frequently consider additional factors. Just to identify the "social indicators" often considered, for example, Tibor Scitovsky measures life expectancy at birth, infant mortality per 1000 live births, daily calorie intake per capita, daily protein intake per capita, residential floor space per capita, percentage of households with running water, percentage of households with television sets, percentage of households with passenger cars and/or motorcycles, electric power consumption per capita, and indices of income inequality (1985, 216–217).

While there have been major critiques of Rostow's historical analyses, there is little doubt about its ideological importance as a guide to international economic development policies through the second half of the twentieth century, in part because Rostow served as the chief architect of American policies regarding economic development from the Nixon through the Reagan administrations, and in part because when he was not serving in the government he was a leading theorist of development. Though he was now teaching at the University of Texas, his views dominated Ivy League economics departments where many of the generation of technocratic economists who guided the late-twentieth-century development efforts of such nations as Mexico, Singapore, South Korea, and Brazil were trained.

Not only did Rostow start from the unanalyzed assumption that membership in the world market system should be the goal of economic development, he used the ultimate Taylorite efficiency measure—that is, the average productivity per labor hour—to define the stages of economic growth. According to Rostow, all economies began as subsistence economies dependent on some form of agriculture or herding. At this stage, in which family units were the sole units of economic production, productivity per labor hour was so low that it did not allow for any capital accumulation. In the next stage of "preparation for take off," some kind of green revolution occurred so that agricultural production per unit of effort increased beyond the subsistence level. Surplus produce could then be sold in a local, regional, or global market so that a small amount of capital could be accumulated, usually on the order of 5 percent of the total production per annum (Rostow 1990, 44). Next, in the period of economic "take off," if people were willing to forgo spending to immediately improve their standard of living, the capital accumulated through agricultural activity could be invested in entrepreneurial enterprises—sometimes in commercial agriculture but often in industrial activities—which incorporated even more efficient production, yielding annual capital growth on the order of 7–10 percent (Rostow 1990, 44–45). A self-denying population bent on producing for export markets could enter a "drive to maturity" stage of continuing productivity enhancement and capital accumulation which could eventuate in a "mature" industrial economy like those of the Britain, the United States, France, and Germany in the twentieth century. According to Rostow, Britain had entered its take off stage around 1783, The United States around 1840, and Germany around 1850 (Rostow 1990, xviii).

Several implications for nations seeking economic development seemed to follow from Rostow's model. First, the closer a nation was to a subsistence economy, the more important it was to invest initially in agricultural improvement to create a commercial agriculture that could generate capital. Conveniently for the industrialized nations, this perpetuated the long-standing demand for commercially oriented monocropping within developing economies. Moreover, this strategy was strongly promoted simultaneously by the proponents of a "green revolution," including such organizations as the Rockefeller Foundation, whose avowed interest was in ensuring an adequate food supply for a growing world population (Glaeser 1987).

Unfortunately, the green revolution was dependent on the selection of high yield crops which demanded very rich soils and/or huge amounts of fertilizer and pesticides. This situation meant that only farmers who already had substantial available capital could benefit from producing the new high yield crops, creating a major pattern of income concentration within agriculture and driving many poorer peasant families out of agricultural production and into urban poverty (Villianatos 2006, 59). It also encouraged the longer term degeneration of soils, which led to decreased yields after a short period of high level production. In a sense, then, the Rostow model, like other technocratic models, was willing to overlook issues of income distribution in favor of overall productivity increases, and it was relatively insensitive to issues of long-term sustainability.

Second, because every step in improving productivity per labor hour—even in relatively labor intensive industries—called for increased capital investment, newly developing countries needed to adopt a combination of strategies to generate increasing internal capital investment. One strategy utilized by virtually all successfully developing economies in Asia and Latin America was to directly or indirectly discourage the production of consumer goods for the home market so as to direct increased productivity into capital investment in industries that had significant multiplier effects, such as the production of machine tools and equipment, or into investment in production for external markets which yielded increased capital. In South Korea, for example, largely as a result of governmental export incentives to be discussed below, while GDP increased by 8.6 percent per annum between 1965 and 1981, net exports (the value of exports less the value of imports needed to produce them) increased by 26.0 percent per annum (Scitovsky 1985, 216). Such strategies were viable almost exclusively under authoritarian regimes, because where democracy prevailed it was difficult to enforce the kind of delayed gratification which they demanded (Centeno 1994, 28–29; Muller 1985, 446–447).

Where capital accumulated in private hands, developing economies also had to find ways to restrict the export of internally generated capital in order to be successful because wealthy persons, being risk aversive, were naturally inclined to invest in stable economies, especially those of the industrialized West. So in spite of Rostow's promotion of free market economics, his model demanded severe governmental restrictions on capital flow. Even when nations sought to invest in education to increase "human capital," since higher education was initially available only in the developed nations, developing nations often attempted to institute restrictive policies to prevent the "brain drain" of educated persons who saw greater personal opportunities if they remained in the developed nations where they finished their education rather than returning home (Haque and Kim 1995, 577–581). Far from promoting democracy, then, American policies based on the technocratic Rostow model prioritized economic growth over the promotion of democratic political institutions or even certain aspects of free market capitalism during the last third of the twentieth century (Muller 1985, 447–448).

Like Taylorism, while the model was intended to serve the general welfare rather than just the interests of the capitalist and/or educated classes, it often did favor the latter. Throughout much of India, for example, green revolution technologies favored those wealthy peasants who had money to advance for fertilizers and pesticides, reducing the income of poorer peasants (Glaeser 1987, 57—75). The guiding principle, like that of Taylorism, was increase production first, then worry about the distribution of increased wealth.

Because the internal accumulation of capital was inevitably slow, all developing economies except that of Taiwan sought to attract international capital as well, which demanded that they establish a significant amount of trust on the part of international bankers and other investors and that they offer opportunities for relatively high yields on investment. Finally, just as they had to ensure that locally generated capital did not leave the country, they had to find ways to induce international capitalists to re-invest substantial amounts locally. All of these considerations demanded that those developing countries which had nationalized major industries privatize them once again in order to ensure foreign investors that their capital would not simply be co-opted by the state. Furthermore it demanded that the developing nation's economy be firmly under the control of economists with specialized training—that is, technocrats—who could be trusted to understand and operate by the same rules as foreign investors (Centeno 1994, 24–25).

The Aesthetics of Development: High Modernism and Its Problems

Though the concept of *High Modernity* as opposed to mere modernity was used at least as early as 1992 by Daniel Hallin (14) it became widely used in connection with notions of development through the publication of James C. Scott's *Seeing Like a State: How Certain Schemes to Improve the Human Condition Have Failed* in 1998. According to Scott, the High-Modernist ideology which dominated in development circles from the mid-1950s through most of the 1970s,

> is best conceived as a strong, one might even say, muscle bound, version of the self confidence about scientific and technical progress, the expansion of production, the growing satisfaction of human needs, the mastery of nature (including human nature), and, above all, the rational design of social order commensurate with the scientific understanding of natural laws. . . . It was, as the term "ideology" implies, a faith that borrowed, as it were, the legitimacy of science and technology. . . . The carriers of high modernism tended to see rational order in remarkably visual aesthetic terms. For them, an efficient, rationally planned city, village, or farm was a city that *looked* regimented and orderly in a geometrical sense. . . . They tended to prefer certain forms of social organization (such as huge dams, centralized communication and transportation hubs, large factories and farms, and grid cities). (Scott 1998, 4–5)

Finally, and perhaps most importantly, high-modernism focused on scientific claims to universal knowledge and tended to ignore "the role of local knowledge and know-how [what Scott called *mētis*] which can come only from practical experience" (Scott 1998, 6).

Another set of terms used by Scott to characterize high modernism was "transparency" and "legibility." For the Logical Positivists of the 1930s, transparency and legibility described the ideal characteristics of a scientific claim: it should be unambiguous, univocal, and immediately obvious to anyone who understood its meaning. As applied to high-modernist organizations and structures, transparency or legibility meant that the appearance of any structure or organization should clearly and unambiguously reflect its function, and this implied a preference for structures and organizations which had simple and single uses, rather than complex and multiple uses. Le Corbusier, whose vision defined high-modernist architecture, wrote,

> the inadequacy of traditional techniques have brought in their wake a confusion of powers, *an artificial mingling of functions* only indifferently related to

> one another. . . . We must find and apply new methods . . . lending themselves naturally to standardization, industrialization, Taylorization. . . . if we persist in the present methods by which two functions are mingled and interdependent, then we will remain petrified in the same immobility. (In Scott 1998, 109)

One consequence of the high modernist obsession with simplicity, according to Scott is that, "it always ignores essential features of any real, functioning social order" (Scott 1998, 6).

In agriculture, for example, high-modernism preferred monocropping with rows that were either straight or which followed isoclines over traditional peasant planting systems, which looked to be based on more or less randomly placed hummocks in which multiple species—that is, corn or rice, beans, and yams and pumpkins—were planted together. Since the high-modernist view reinforced the emphasis on cash cropping which was encouraged by the interests of the developed nations, it was doubly powerful. It has generally increased the production of whatever commercial crop was the focus of development experts, but Evaggelos Vallianatos has pointed out one of the major problems with this view:

> Chiapas peasants, who are fighting for survival, raise two tons of maize per hectare while the industrialized farmer next door produces six tons of maize per hectare. For this reason, the agricultural experts call the peasants backward and insist that they leave the land or adopt the methods of the mechanical plantation. Yet the industrialized farmer gets nothing more from his land but the six tons of corn. The Chiapas peasant, however, grows not only maize, but along with maize, he raises beans, squash and pumpkins, sweet potatoes, tomatoes and other vegetables and fruits and medicinal herbs. Some of his food the peasant sells for cash and the rest is for his family, chickens and cattle. The Chiapas peasant easily produces more than fifteen tons of food per hectare and all without commercial fertilizers or pesticides and no assistance from banks or governments or transnational corporations. (Vallianatos 2006, 273)

The desire for transparency and simplicity embodied in the drive for maximum production in agriculture and for monocropping has also led to a substantial reduction in genetic diversity among commercial crops with several negative consequences. Sometimes, as was true of traditional tomato varieties and Freestone peaches, taste and texture have been sacrificed in the search for varieties that do not bruise as easily and which can thus be mechanically harvested with minimal loss (Scott 1998, 267–268). More importantly, if all plants in a field are genetically identical they are all susceptible to the same pathogens, and this can lead to disastrous epidemics. In some cases it has been relatively easy to develop pesticides to protect such crops; but the use of pesticides leads to the development of resistant strains of pathogens and to the need to use higher concentrations of pesticides or the need to create new pesticides which starts a new cycle of resistance, and so forth. (Scott 1998, 268–269)

At least into the 1980s, expert agronomists tended to depend upon agricultural knowledge gained primarily in the United States and Europe where topsoils are generally relatively heavy and deep. But that knowledge caused major problems in Africa where topsoils tend to be relatively light and shallow. By encouraging plowing to prepare for planting, agronomists virtually guaranteed that the African soil would be carried away by water or wind with a commensurate decline in agricultural productivity (Bass 1997, 119–121). Familiarity with successful irrigation in Europe and the American West also led high-modernist advocates of development to place a huge emphasis on irrigation schemes based on physical structures without much concern for social structures. In discussing the Niger floodplain (the Sahel) in Mali, for example, Thomas Bass, a journalist who has spent much of his life in Africa wrote, "A Western engineer . . . sees a resource waiting to be exploited, a good place to build dams and other projects cast in concrete." But there are two problems with this view: first, "the soils are too fragile to support the irrigation projects already built in the Sahel. Most of these have been plagued by waterlogging and salinization before finally being abandoned." Equally importantly, "Permanent waterworks require social hierarchies to run them: managers on top, workers below. But the network of relationships in the Sahel tends to stretch horizontally among peers rather than vertically between figures of authority" (Bass 1997, 21–22).

Bass has also described the World Bank-funded International Institute of Tropical Agriculture (IITA) near Ibidan Nigeria in a way that calls attention to the insensitivity of high-modernist agricultural experts in the 1960s and 1970s to local geological and cultural conditions—an insensitivity which led local Nigerians to refer to the isolated IITA compound staffed with one hundred and fifty scientists and a thousand assistants as "little America": Educated primarily at land-grant universities in the United States, its scientists often knew more about farming in Iowa than in Africa. They relied heavily on fertilizers and other chemical interventions [which peasant farmers could not have afforded even if they had been effective]. They were embarrassed by "improved" rice strains that no one wanted to eat [at least in part because they were a different color than traditional local strains] and mechanized agricultural schemes that failed one after another. Although women grow much of the continent's food, the experts and their new technology seldom reached them. Also forgotten was the troubling fact that an increase in food production does nothing to solve the inequities in who gets to eat the food (Bass 1997, 122–123).

Since the late 1970s high-modernist agriculture has gradually been replaced by approaches much more attuned to local circumstances; but with rare exceptions those approaches have continued an emphasis on cash cropping. The IITA, for example, has reduced its emphasis on high yields in favor of plants resistant to local and regional diseases, pests, and fluctuations in rainfall, it has focused more on sustainability under local circumstances, and there has been greater emphasis on genetic diversity (Bass 1997, 123–124).

The high-modernist preferences for transparency and regularity as well as largeness are literally most visible in connection with village and city planning both in developing countries and in rural regions of developed countries. Brasilia, the Brazilian capital, designed by Oscar Niemeyer, one of Le Corbusier's students, along with Lucio Costa and completed in 1960 after just four years of construction, is the paradigmatic example of high-modernist planning in the developing world, while the cities of Frobisher Bay and Inuvik in the Canadian arctic are examples of high-modernist planning in a developed nation.

Located on a plateau over 1000 kilometers from the major population and commercial centers of Rio de Janeiro and Sao Paulo, Brasília offered the opportunity for planners to start completely fresh, without considering either local or national customs, habits, or land ownership patterns. Its plan would be completely rational and would create a new, standardized, citizenry for the future (Scott 1998, 118–119). Its basic layout was that of an arc of residential zones called "superquadra" bisected by a central axis containing a "plaza of the three powers" and an "esplanade of the ministries" on a mammoth scale.

Since the city was designed as an administrative center, the esplanade of the ministries constituted the "work" region of the city, which was clearly separated from the residential region. Moreover, each ministry had its own separate building.

Housing was organized through "superquadra." Each superquadrum was a cubic building containing 360 nearly identical apartments housing approximately 2000 residents with its own nursery and elementary school, while each grouping of four superquadra included a retail sector, a secondary school, a sports facility, a social club, and a cinema (Scott 1998, 125). Furthermore, each superquadrum housed only people associated with one ministry or profession. Finally, the superquadra groupings were linked to the administrative center of the city and to the external world by a wide central arc boulevard designed to carry only vehicular traffic, so there were no paths for pedestrian traffic among the superquadra groupings.

All of the well-defined formal needs of the population for food, clothing, housing, education, entertainment, and exercise could be met within a single grouping of superquadra; but residents, who came from more traditional cities and towns, were used to living in small heterogeneous neighborhoods in which street corners and small squares provided opportunities for informal conversations and public events, such as religious celebrations, farmers' markets, and so forth, and those activities contributed critically to a satisfying sense of community life. Most residents thus found Brasília an alienating

Figure 5.1. The "esplanade of the ministries" in Brasilia. Each side of the central open space mall contains an identical set of buildings, each of which houses a separate governmental ministry in conformity with the high modernist notions that architecture should be monumental, that spaces should reflect geometrical regularity, and that each space should contain a single function. Source: Photograph by Vanessa Volk.

environment. The architects also failed to account for housing the construction workers and the many service industry people who would run the city; so an unplanned region of housing for the underclasses grew up outside the planned region. This disparity in housing heightened class differences in a city that was supposed to be egalitarian (Holston 2009, 93–103).

During the Cold War, the creation of the Distant Early Warning Line (the "DEW Line") of radar sites in the Canadian North stimulated interest in the Arctic and attempts "to modernize the North through new techniques and technologies [in] the belief that the distinctiveness of northern landscapes could be subdued or even overcome" (Farish and Lackenbauer 2009, 520). This impetus was encouraged by a CIA financed research Center for International Studies (CENIS) at MIT where Walter W. Rostow had been investigating Soviet society before moving into the Kennedy administration. According to CENIS scholars, villagers in general, including those of the Canadian North were "sedentary . . . apathetic, and resentful," and the goal of development was to transfer, "the aesthetic of modernist [read high-modernist] urban planning to the village[s]" in order to turn their residents into model modern citizens (Farish and Lackenbauer 2009, 523). Though an Advisory Committee on Northern Development (ACND) had existed since 1948, it was only after the Department of Northern Affairs and National Resources (NANR) was created in 1953 that the ACND began to guide the attempt, "to enable [northerners] to share fully the national life of Canada" (Farish and Lackenbauer 2009, 526). Furthermore, because of the strong military presence associated with DEW Line activities, many specific development projects were driven by military considerations, with their characteristic quest for clarity and simplicity.

The area around Frobisher Bay had been part of a hunting and trapping region in which the Inuit of southern Baffin Island had several seasonal and permanent camps for nearly a century, but it was described by one urban Canadian as "nothing but a barren valley" in 1957, when plans were being made to turn it into a "model Arctic community." It would serve as a transshipment point for military materiel traveling both to Europe and to the eastern portions of the DEW Line as well as a halfway refueling point for commercial flights from the United States to Europe (in Farish and Lackenbauer 2009, 528, 530).

A military air base had existed there since the mid-1940s. That had been vastly expanded in the early 1950s by the U.S. Air Force. The airbase was turned over to the civilian Department of Transportation in 1957, and by 1958, Pan American Airlines was using it as a stop-over for passenger service to Europe. Moreover the Hudson's Bay Company established a post nearby to take advantage of the newly available air transportation, and Frobisher Bay seemed poised to become a full-fledged town.

Up to this point, housing for Inuit construction workers had not been provided by the U.S. military; so a makeshift shanty town surrounded the air base. But in 1958 the Canadian government decided to make Frobisher Bay a planned city to house "some thousands of people, and their recreational and even work places in a completely integrated complex." According to an account in *The Canadian Architect*, it would have "a tightly-planned town center totally closed within a dome and surrounded by 36 apartment towers of twelve stories each," and according to *Time* magazine it would be, "a dream town [that might] prove the prototype for next-century towns across the North" (Farish and Lackenbauer 2009, 530). Unfortunately, technology and military events overtook the grand plans for Frobisher Bay, as mid-flight stopovers for intercontinental flights became unnecessary and as the U.S. military presence in Northern Canada declined during the early 1960s.

Planning was down-scaled, but Frobisher Bay did become an administrative center for Northeastern Canadian business enterprises. The residential area around the HBC post was planned down to the paint schemes for the houses, and a school, a nursing facility, and a Government Rehabilitation Center to retrain Inuits for wage labor were constructed. In addition, the government did build some seven-story apartment towers near the air base. Without the military presence and transportation hub business, however, the number of wage labor jobs available nearby was very small. The Inuit population which had been attracted away from their traditional hunting camps by the promise of modern amenities and opportunities, were mostly unemployed, and Frobisher Bay (renamed Iqualit in 1987) developed a reputation for "drunkenness, violence and prostitution," which led Northern whites to view it as "the embodiment of all that is wrong in Northern development" (Farish and Llackenbauer 2009, 531).

Inuvik, which became the administrative center for the western part of the Canadian North offered a slightly different pattern of high-modern development. Western trade and administration had been located in the village of Aklavik during the early twentieth century; but its location made Aklavik subject to frequent flooding and continuing erosion. So in 1953 the ACND in Ottowa recommended that all government activities be moved to a new site that would be "consciously designed to demonstrate the possibility of building a northern town with as many of the features of our urban civilization as possible." Substantial incentives were offered to those natives who would move from Aklavik to the new site, named Inuvik, and about half of them did move (Farish and Llackenbauer 2009, 533–534). Chosen by engineers to be on a flat, dry, area to the east of the MacKenzie River which was so short of fish, game and fur resources that neither the Cree nor the Inuit hunted there, the new site might have been suitable for an administrative center, but it was not suited to the indigenous population of Aklavik, who had continued to supplement any urban wages with hunting and fishing income.

Unlike Frobisher Bay, Inuvik was a totally planned community with a modern school, an eighty-bed hospital, and modern and fully equipped houses and apartments with central heating and an above-permafrost water and sewage system—for those who relocated from the South. The indigenous population was concentrated in an "unserviced" residential area, however, where standardized housing was a 512 sq. foot prefabricated house with neither water nor sewer service. According to the anthropologist Alexander Ervin, among the planners from Ottowa,

> There had been a pervasive faith in the positive effects of modernization and economic development, but this mega-project, the building of a new town and the relocation of a whole community, was done primarily for engineering and economic reasons rather than being based on carefully thought-out human needs or consequences. There was virtually no consultation with the people most affected, and the decision makers were distant, both geographically and culturally. (In Farish and Lackenbauer 2009, 538)

Without any consideration for the culture of those who would be the primary inhabitants of the city or of the regional natural resources that could provide the foundation for long-term employment in the region or the skills that employment might demand, the one-size-fits-all ideology of high modernism was transported to the Arctic. Moreover, since appropriately modern education was taken by development experts to be critically important, family welfare allowances were withheld from Inuit families who did not send their children to the schools which existed only in the large settlements like Frobisher Bay and Inuvik. By standardizing curricula and moving students to these residential schools, according to C. W. Hobart, an expert observer, the Canadian government, "made children unfit to live in the land camps which are their homes [without making them] adequately prepared to make a successful transition to wage work in town" (In Farish and Lackenbauer 2009, 524). Thus high-modernist sites like Frobisher Bay and Inuvik tended to undermine the economies of large surrounding regions.

Examples of Economic Development, South Korea and Mexico: Different Paths with Different Consequences

Development efforts in different countries have had drastically different consequences. Among the so-called BRICK countries, (Brazil, Russia, India, China, and Korea) economic growth and development have been both significant and sustained over substantial periods of time. Among many others, especially Mexico and several Latin American countries including Peru and Venezuela, progress has occurred in fits and starts with brief periods of growth punctuated by longer periods of stagnation or decline. And among still others, especially in Africa, development efforts have been largely ineffectual, with populations growing faster than GNP. We have already considered some of the conditions that have made economic development—especially agricultural development—in much of Africa so problematic; so we look here at one example from each of the other two categories, hoping to identify general considerations which differentiate between the successes and the "its not clear yets."

At the end of the Korean war, South Korea had a number of critical advantages that made development relatively easy. Some, though not all, were shared with other "Asian Dragons," including Taiwan and Singapore. First, and ironically, because the Japanese conquerors during WWII had expected to retain control, they had invested substantially. The Japanese introduced new high yield rice and established agricultural research stations to improve agricultural production, they built roads, railroads, and harbors, and they began to establish industrial plants. When WWII ended, Japanese-owned factories and land were confiscated and the land was distributed among small peasant farmers. The Korean government was also strong and autonomous enough to force large Korean landowners to sell land to small farmers well below market rates, thus reducing inequalities of wealth and creating a strong domestic market for goods and services. Only Taiwan among developing nations during the second half of the twentieth century was more egalitarian with regard to wealth, and because Koreans spent more of their earnings on private consumption, their domestic market was at least as strong as that of Taiwan (Scitovsky 1985, 217–218). Farm price subsidies also served to increase rural earnings and stave off a larger immigration into cities than increased industrial and commercial employment opportunities warranted (Scitovsky 1985, 220).

Korea also shared with the other Asian dragons a culture which regards learning very highly, and while literacy was low—13.4 percent—at the end of WWII, an emphasis on education led rapidly to a virtual elimination of illiteracy. Korea offered six years of compulsory schooling for all, and family expenditures for additional education brought total educational expenditures on education up to 9 percent of GDP by 1985. More than 75 percent of the population had attended secondary schools and nearly 12.5 percent had attended college in 1985 (Scitovsky 1985, 219). By encouraging the development of industries such as the electronics industry, which demanded relatively high levels of education, Korea managed to keep a very substantial fraction of those who represented high levels of human capital at home.

A long-standing East Asian custom of augmenting worker income by paying end of year bonuses allowed and encouraged a relatively high personal savings rate—around 7.6 percent of disposable income—in Korea (Scitovsky 1985, 243). In addition, the Korean government was able to make capital export illegal (Scitovsky 1985, 253). As a consequence, more than 2/3 of the capital used to fund productive industries and economic growth in Korea between 1965 and 1981 came from domestic sources, reducing the need for foreign capital and limiting the transfer of capital out of the country for debt service on foreign loans. Relative to Taiwan, which was able to finance roughly equal economic growth without any foreign capital, Korea might seem unsuccessful, but by any other standard the performance of the Korean economy and its ability to provide social services was quite remarkable.

Finally, the Korean government developed a series of policies to encourage industries that produced for export, to increase employment, and to increase efficiency in terms of GDP per capita. These policies included tax concessions, credit on specially favored terms, and a variety of informal pressures. Because the government owned the Bank of Korea, it was able to make very low interest loans to selected industries, and in order to appeal to some foreign capital, the government paid the difference between the going rates on foreign loans and the favorable rates it promoted to encourage development. The government balanced carefully between encouraging labor intensive and capital intensive industries so as to keep employment strong while increasing efficiency, and it balanced between industries for the domestic market (which it often protected through tariffs on imports while the industry developed the ability to compete in global markets) and export industries which could be competitive. Wage rates were controlled so as to keep inflation tolerable.

All of these policies were made possible because, though nominally democratic and civilian, the Park regime had the support of the military and faced neither leftist nor more conservative opposition. From 1961 through 1979 it was in effect an autocratic regime. In addition, it was a regime dominated by technocrats within virtually all ministries. These technocrats were able to engage in well-coordinated planning and policy implementation. As a consequence, between 1965 and 1981, unemployment averaged less than 3.4 percent, annual GDP grew by 8.6 percent annum, exports grew by 26 percent per annum, labor productivity in GNP per employed person grew by 5.2 percent per annum, real wages in manufacturing industries grew by 7.9 percent per annum, and real consumer spending grew by 5.5 percent per annum, while income concentration continued to decline slowly (Scitovsky 1985, 216).

At the beginning of the 1990s under the presidential leadership of Carlos Salinas, Mexico looked to be achieving economic results comparable to those achieved earlier by South Korea. Inflation, which had been 159 percent per annum in 1987, had been reduced to 12 percent by 1992. Public sector deficit, which had been 17 percent of GDP in 1987, had been turned into a surplus by 1991. GDP was growing by 3.9 percent per annum, and private investment had increased by 25 percent during Salinas's presidency (Centeno 1994, 16). Many Russians and East Europeans looked to Mexico and the policies of Salinas's government as models to emulate under their new regimes (Centeno 1994, 231). But in 1994 Mexico saw a huge reversal of fortunes. Leftist revolt in the state of Chiapas and the assassination of the dominant Institutional Revolutionary Party's (PRI) presidential candidate and Salinas's close friend, José Ruiz Massieu, demonstrated the fragility of the stability of the Mexican government and led to a crushing withdrawal of foreign investment. In 1995 the value of the peso dropped 50 percent, and debts, which were largely stated in U.S. dollars, bankrupted thousands of businesses, creating an economic collapse.

In order to understand both the apparent success of Mexican economic growth during the 1980s and its reversal in the 1990s we need to explore the conditions governing the growth and dominance of a technocratic state in Mexico beginning in the 1970s. Since 1940, Mexico had essentially been a one-party state. Though nominally a democracy, the *Partido Revolucionario Institucional* effectively controlled the state, and political negotiations among various interest groups were largely handled within the party. Within the government/party, the president was extremely powerful. Though he could not be reelected after his six-year term, the president basically named his successor, appointed cabinet level federal personnel, and named the party's state governor candidates.

Under the president until 1970 the two most powerful ministries were the Ministry of Government and the Ministry of the Treasury. The former dominated the government; it controlled spending after the budget had allocated specific amounts to the various ministries; and it managed the patronage system by determining just who got to spend what on specific programs and projects. The president always came from the Ministry of Government. The Ministry of the Treasury was the second most powerful ministry, charged with establishing the basic budget, managing fiscal policy, public debt, and economic policy. It was staffed at the top largely by economists and managers who had finished their education abroad, usually in the United States or in Europe.

Beginning in 1970 under the leadership of President Louis Echeverria, two major innovations began. First, the planning function was elevated in importance and it began to be dominated by the PRI's think tank, the *Institutio des Estudios Politicos, Economicos y Socials*, (IEPES) whose members were technocrats largely insulated from popular political concerns. These men and women tended to move into leadership positions within the government. In fact, by 1983, over 50 percent of the Cabinet had been IEPES members (Centeno 1994, 84). Secondly, the government bureaucracy grew rapidly from just over 600,000 employees in 1970 to 2.1 million in 1976 to 3.3 million in 1983. This expansion was paid for by massive public borrowing and by oil revenues (Centeno 1994, 82). At the upper level virtually all of these new positions were filled by technically trained personnel in agronomy, engineering, management, and economics. At the same time, cabinet level appointments were increasingly dominated by foreign educated management and economics-trained technocrats.

Under Echeverria roughly 25 percent of all cabinet members had graduate degrees in economics or management from foreign universities, and this increased to over 50 percent under Salinas, who became president in 1988 (Centeno 1994, 121). As a result, the home grown politicians who had dominated party politics were gradually displaced in power by a technocratic elite, symbolized by the fact that beginning in 1976, with the election of José López Portillo, the powerful presidency was filled from the technocratic rather than from the political branch of the party. Portillo instituted a National Planning System organized by the Harvard educated economist, Carlos Salinas. The National Planning System was intended to be the "instrument to define objectives, proposals, means, resources, and actions with the goal of reaching national objectives of economic and social development" (Salinas in Centeno 1994, 85). Moreover, as part of the new planning system Salinas instituted the Planned Program Budgeting System (PPBS) from the United

States as the new Mexican budgeting system to replace the old system of allocation by administrative unit.

The new technocrats basically ignored conflicts between interest groups within the party in order to institute scientific planning by economists. Thus López Portillo argued, "Economics is [the] science which allows us to conduct and guide the transformation of society" (in Centeno 1994, 217). Perhaps even more importantly, the science of economics which dominated the Mexican government was the neoclassical economics taught at Harvard, Chicago, and Stanford rather than the Marxist version taught at the University of Mexico. Cabinet members labeled leftist critiques as "absurd infantilism," characterizing one leftist critic as someone for whom, "the sun continues to move around the earth" (In Centeno 1994, 221). Moreover, the "scientific" economic doctrines were wielded by economists who had been brought up almost universally within the small population of wealthy Mexican families. By 1989, out of 1,113 mid-level federal bureaucrats only eight indicated that they had come from peasant families and ten that they had come from working-class families (Centeno 1994, 112–113). The economists, who came from the capitalist class, tended to view the freedom of capital as central to economic efficiency; so unlike their Asian counterparts, they were unwilling to place restrictions on capital flight.

Echeverria had attempted to retain some of the revolutionary claims of the PRI by arguing that growth should be combined with economic justice. Thus he offered more social support for the poorest segments of the population. But he had attempted to finance growing public expenditures primarily by foreign borrowing and by nationalizing a number of industries. The number of state-owned enterprises grew from 84 in 1970 to 845 in 1976. But income failed to meet expenditures, public debt quadrupled during the six years of Echeverria's presidency, and there was massive capital flight, in part because of international capital's fears of nationalization.

The message received by López Portillo and those who followed was that government policies could not succeed without the support of domestic and international capital; so they initiated policies that assumed, as had earlier Scientific Management, that economic growth had to precede concerns with distribution. In the long run, Portillo argued, "the worst policy is to turn the economy into a Utopia. Populism does not solve anything, but complicates and enlarges problems" (In Centeno 1994, 184). As advocates of Rostow's model of economic growth, Mexico's technocrats believed that someone would have to make short-term sacrifices to allow for long-term growth, and they likewise believed that realistically, that had to be Mexico's poorest classes rather than the most affluent, capitalist class. Thus, government expenditures for economic development grew much faster than welfare budgets.

For a short time income from oil and relatively low interest rate foreign loans kept income and expenditures balanced and allowed for modest economic growth, but a decline in oil prices and increase in interest rates on loans starting in 1981 led to a drop in GNP, a new flight of internal capital, rampant inflation, and increasing public expenditures to meet debt service which had been running at about 32 percent of the budget up to 1981 but jumped to over 45 percent in 1982, reaching over 60 percent by 1989. To meet the new crisis the government sought to sell off state-owned firms and drastically reduce public expenditures, which shifted virtually all costs to the bottom 2/3 of the population (Centeno 1994, 192). The social services portion of the budget was reduced from roughly 30 percent in 1981 to about 20 percent in 1982, and eventually to 15 percent in 1989 (Centeno 1994, 199).

The result was that income concentration continued at an accelerated rate, and by 1990 average per capita meat consumption had declined 50 percent from 1980 levels, milk consumption had declined 40 percent from 1970 levels, and caloric intake was less than 50 percent of that recommended by the World Health Organization (Centeno 1994, 206–207). Between 1988 and 1991 the social welfare budget was increased just enough to allow the regime to remain stable; but leftist opposition was increasing, and by 1995 even electoral fraud and violence could not stave off collapse.

Mexico had started off its economic development push with a number of handicaps relative to the Asian dragons which it hoped to emulate. Its rural population was much poorer, there was virtually no tradition of private savings among peasants or urban workers, nor was there a tradition of education that reached below the top 15–20 percent of the population. This situation made Mexican economic growth more dependent on foreign capital, and it made the creation of internal markets much more difficult. The policies of the technocrats, based as they were on a particular neo-classical model of economic growth, tended to exacerbate rather than mitigate these circumstances. Their willingness to allow domestic capital to leave the country, their failure to enact policies that promoted the reinvestment of profits from foreign investments in the local economy, and their investment of up to 72 percent of foreign loan monies in the stock and money markets rather than in direct productive capacity undermined economic growth (Centeno 1994, 250). Furthermore, their enactment of policies that shifted virtually all cost of growth to the most economically disadvantaged 2/3 of the population not only precluded the growth of internal markets, but it eventually generated so much leftist opposition that it destabilized the government.

Subsequently, Mexico recovered from its 1995 economic collapse to some extent. Northward migration provides a safety valve to limit internal leftist activity, but Mexico's dependence on the persistence of cheap labor to maintain a place in the international market economy continues to preclude a large increase in the income of the working population, and its promotion of agribusiness has essentially destroyed whatever self-subsisting rural peasantry had previously existed. As a consequence there has been little opportunity to decrease income inequalities to create the kind of internal markets that characterize the highly industrialized countries of the world.

China under and after Mao

After the Communist defeat of Nationalist forces, starting in1949 the Chinese Communist Party (CCP) began to look to Soviet experience—especially official accounts of the fourth Five Year Plan for 1946–50—to guide its move toward industrialization under socialism, in large part because the CCP saw their situation in 1949 as similar to that of the Soviets in 1946. They were coming out of a war that had depleted both industrial plant and industrial workers, and they needed to recover economic strength as fast as possible.

Under the Russian plan, Party members, through the Labor Union, rather than independent expert managers, were put in charge of promoting "the struggle for plan fulfillment." The specific techniques they were to use involved a combination of compulsory education in the goals of the plan and the role of science in the struggle, along with strategies used by the Stakhanovite revivers of Taylorism (Bailes 1978, 316–318). Henceforth, technical expert management of socialist factories—preferably by educated proletarians—was understood as a necessary stage in the transformation from Socialism to full Communism. Under this plan, labor discipline was extremely strict, and workers could be imprisoned for a variety of minor infractions, including being late for work or failure to enthusiastically adopt the practices of "rationalizers"—teams of highly skilled workers who demonstrated new and more efficient production practices. "Socialist competitions" to increase efficiency were emphasized (Kaple 1994, 28–32).

The fact of the matter was that the Soviet fourth Five Year Plan was ineffective and was dropped by the Soviets in 1947; but this situation was unknown to the CCP, which went forward with increasing both Party responsibility for improving industrial productivity and increasing labor discipline. These strategies did have the effect of increasing Party power; but they had little more success in increasing industrial efficiency in China than they had in the Soviet Union; so Mao Zedong promoted the Great Leap Forward in 1958. During this new push to increase both agricultural and industrial productivity, Taylorite scientific managers from the Soviet Union came to China to help increase industrial production with little positive effect. Mao formally accepted responsibility for the failures of the Great Leap Forward and accepted his own removal from the presidency, but he also attacked both Khruschev and the Soviet engineers as "Soviet revisionists," leading to a long-term split between the two greatest Communist regimes.

Briefly, from about 1962 to 1966, moderates, including Deng Xiaoping, began to gain political power in China. Deng had spent time during the early1920s in France, working first at the Le Creusot Iron and Steel plant in central France then as a fitter in the Renault factory in the Paris suburb of Billancourt where Scientific Management practices were implemented. Deng would later claim that it was here that he was introduced to Scientific Management from the workers' point of view and got an initial feel for the evils of capitalist society. He studied Marxism and joined the French communist party, then studied in the Soviet Union until 1927. There he picked up the Leninist view that while it was exploitative under capitalism, Scientific Management might be valuable under socialism. Though Deng received a thorough grounding in Marxism-Leninism in France and in Moscow, his Marxism was vastly more flexible than that of most early Chinese communists. In particular, like Lenin in the early days of Soviet rule, Deng was sympathetic to technical experts and convinced that they would have to play a key role in China's industrialization and modernization.

From 1927 to 1949 Deng had been involved in Chinese Communist warfare against the Nationalists, against Japan, and against the Nationalists again. A participant in the 1949 fighting, Deng took on increasingly important political roles, beginning as mayor of Chongqing. By 1956, he had become the General Secretary of the Chinese Communist Party (CCP) Central Committee, Director of the Organization Department and Vice Chairman of the Central Military Commission.

In 1966 Mao initiated the Cultural Revolution to restore his own power. Aimed against the revisionism associated with the Soviet Union under Khruschev and against intellectual expertise in general, the cultural revolution stripped Deng along with other moderates of all power and exiled him to the countryside, where he worked in a tractor factory for four years.

By the early 1970s the cultural revolution had ended and in 1974 Premier Zhou Enlai returned Deng to power as first Vice Premier in charge of renewing the economy after the Cultural Revolution. When Zhou died, power briefly fell into the hands of the ultra leftist Gang of Four, and Deng was stripped of power again. But when Mao died in September of the same year, Deng gradually resumed power, and on July 22, 1977, Deng was restored to the posts of Vice-Chairman of the Central Committee, Vice-Chairman of the Military Commission and Chief of the General Staff of the People's Liberation Army, becoming the *de facto* leader of China.

From his new position, Deng began a process of moving young—in their 30s—engineers into important political positions. Wang Zhaoguo, graduate of the Harbin Institute of Technology in 1966, was made director of the Office of the Central Committee, and Ding Guan'gen, who graduated in engineering from Jiaotong University, became minister of transportation. Two other long-term moderate colleagues of Deng were moved into prominent positions. Hu Yaobang became Party General Secretary and Zhao Ziyang became premier and then party chief. Both took on roles as technocratic patrons following Deng. Hu argued that the CCP should establish the policy that, "all leading Personnel must be trained Specialists" (Li & White 1990, 12), and Zhao established a number of research centers or "think tanks" to advise the government.

In a fascinating comparison of elementary school language textbooks from the early 1970s with those from 1979, Julia Kwong has demonstrated how deeply the shift from Mao's aversion to intellectuals to the Deng generation's admiration for technical experts penetrated Chinese culture. Like language arts texts everywhere, Chinese textbooks contain short stories for students to read, and these stories embed cultural attitudes. Leaving aside direct references to or quotes from political figures, the major change in elementary language texts was who appeared as heros and villains in these stories. In the earlier period the heros were farmers and workers, while the villains were landlords, capitalists, and anyone who might want to promote private enterprise. After 1979, in addition to peasants who entered into free market activities, engineers, scientists, and agronomists were the heroic figures, and children were encouraged to get educations in these fields. Zhan Tianyou, who built the first railroad in China was lionized, and even capitalist technologists, including Thomas Edison, were praised. In one story from 1979 a grandfather asks his grandson about his career plans. The grandson replies: "I would like to be an agricultural scientist. When grandfather heard this, he was very happy, and said, 'My good grandson, would you like to work in your home town?' I returned his glance. He looked so serious, as if making the offer to a specialist. I happily agreed" (Kwong 1985, 202).

Between 1980 and 2007 China's political elite became increasingly well-educated and it came increasingly from persons educated as engineers. While the term, "engineering," has been applied to a wide spectrum of programs with a huge variation in academic rigor in China, recent studies suggest that of the 517,225 engineering graduates from Chinese institutions reported for the 2004–2005 academic year, roughly half graduated from programs equivalent in rigor to U.S. bachelors programs in engineering, which produced 133,854 degree holders (Wadhwa, Cereffi, Rissing, and Ong 2007, 3). By the late 1980s, 75 percent of mayors of Chinese cities had an engineering or science education. Of those, 64 percent had engineering education (Li & Bachman 1989, 75), and most of those had engineering experience before taking on a political appointment (Li and Bachman 1989, 78). Since being a mayor is the typical first step in moving into more important positions in the CCP, by 2007 every member of the Standing Committee of the Politburo of the CCP was either an engineer or a scientist (Suttmeier 2007, 71).

The new engineers have indeed manifested technocratic attitudes—including a surprisingly lax attitude toward income inequality for socialists. It is also clear that the huge Chinese emphasis on infrastructure development, including such massive engineering projects as the Three Gorges Dam and the South-North Water Transfer Project, which began in the late 1980s is a consequence of the engineering orientation of major political figures. So too is the Central Committee's policy of asking technical experts to hold seminars for them before major decisions are made (Suttmeier 2007, 72).

One question that immediately arises is whether the many students—also primarily engineering and science students—who were educated largely overseas have carried Western values, including democratic values, into China. Returned students do hold important positions. In 2006, for example, 32 of 356 regular or alternate members of the Party Central Committee had spent a year or more abroad as students; the majority of presidents of China's universities were returned scholars; 54 percent of the members of the Chinese Academy of Engineering; 81 percent of the members of the Chinese Academy of Sciences had studied abroad; and nine out of the ten most influential economists in China were Western educated (He Li 2006, 13–14). Though some scholars suggest that the returned scholars are associated with the "liberalization" of such institutions as the universities, interviews done in 2004 suggest that returnees, "generally wish for a strong state that would bring vigorous economic reform and incremental political reform. They support a government run by pragmatic technocrats, much like South Korea under the last dictator, Park Chung Hee, and Singapore under strongman Le Kuan Yew" (He Li 2006, 18). If, as I have suggested above, engineering training tends to produce an impatience with democratic processes, that would help to account for the relative lack of interest in democratic reforms on the part of returned students.

It is not completely clear that the rise of engineering trained technocrats in China was so much focused on engineers per-se as it was an artifact of the fact that about 67 percent of all higher education degrees were in engineering during the ramp up of higher education over the past fifty years. The most recent Party congress, in 2007, saw 75 percent of newly appointed provincial leaders drawn from social science and humanities degree holders, and that may represent a new emphasis on social equality, consistent with a growing preference for policies aimed at "a harmonious society" rather than the Deng-promoted and more technocratic policies favoring building, "a comprehensively well off society," without much concern for the distribution of wealth (Young Nam Cho 2008, 158–160).

Chapter Six

Mitigated Scientism and Technocracy

The Evolving Character of Technical Expertise and Advice at the Beginning of the Twenty-First Century

When one reads Frederick Winslow Taylor's *The Principles of Scientific Management* today, it seems remarkable that such a simplistic and insensitive document could have had the kind of extended impact across many cultures that it did. There are a few public intellectuals at the beginning of the twenty-first century, such as Richard Dawkins and Steven Pinker, whose optimism regarding the ability of science to understand everything important to humans and eventually to solve virtually all human problems without recourse to political negotiations, to religious traditions, or to violence remains unlimited. But most experts who seek to influence private or public undertakings now are vastly more sophisticated than Taylor. They are more cognizant of psychological, sociological, and broadly cultural as well as purely economic considerations, more sensitive to the importance of local traditions and values, and more aware that even the notion of efficiency may have different meanings for different communities and depending on whether one is considering the short-term or long-term. No engineering manager, business manager, or public administrator that I know today would argue that there is "one best way" to do anything without regard to its complex context. Moreover, there is a growing movement among elite engineering and management educators, especially in the United States, to broaden the base of technical education to include greater emphasis on social issues and cultural differences (Lucena 2013; Sullivan et al. 2008).

Scientists and technocrats still do often become frustrated when they feel that their well-founded advice is being ignored for "irrational" reasons, whether those are grounded in political, religious, or other sources; and some respond in both very specific and very general ways. One of my close mathematician friends, for example, was recently in charge of a project to advise the Los Angeles County Court system on how to select members of jury pools in order to reduce wasted time on the part of potential jurors and wasted money by the court system in paying for their time. His students developed a mathematical model that could demonstrably reduce wasted time and money, but the judges who commissioned the study refused to adopt it because it could have distorted the age and gender makeup of juries, which he considered a "political" and therefore irrelevant concern. Incensed, he undertook a letter writing campaign to local newspapers denouncing the intentional wastefulness of the court system.

More generally, when their advice is questioned or ignored, some scientists and technical experts complain that those who disagree with them must be scientifically illiterate as well as obscurantists, and that obscurantism, irrationality, and hostility to reason are growing in the world. According to the manifesto of the Italian movement, *Galileo 2001 for Freedom and Dignity of Science,*

> A spectre haunts Italy [read any Western nation], a spectre which provokes alarms by predictions of catastrophe which terrorize the population. It preaches that science and technology are inimical to humankind and nature, and it incites hostility against science by exploiting baseless fears which fog the reason. This spectre is called obscurantism. It assumes various guises, but the most dangerous of them, because of their reactionary nature are environmental fundamentalism [add religious fundamentalism for Anglo-American spokespersons for scientism] and opposition to technical-scientific progress. (In Bucchi 2010, 2)

With rare exceptions these alarmed scientists and technocrats, who constitute a small but vocal minority among scientists, have seen increased science literacy, especially among political leaders, as the principle solution to such problems; because surely if politicians and citizens just understood science they would agree with the experts. Their primary response has been the creation of a "public understanding of science" movement, the most prominent symbols of which have been the establishment of the Journal, *Public Understanding of Science*, in 1992 and a Public Understanding of Science Chair at Oxford University, the first holder of which was Richard Dawkins from 1995 to 2008. In this view, the more extreme advocates of science are simply amplifying a position taken by the AAAS in 1951 when it articulated as one of its major goals—increasing "the public understanding and appreciation of the importance and promise of the methods of science in human progress [because] in our modern society it is absolutely essential that science—the results of science, the spirit of science—be better understood by government officials, by businessmen, indeed by all the people," and by the combined BrASS, Royal Society, and Royal Institution when they established a Committee for the Public Understanding of Science in 1985 (Bucchi 2010, 3–4).

Of course, some cynics have pointed out that at least part of the call for public understanding of science has been self-serving, in the sense that scientists have believed that scientific literacy would generate jobs for scientists. According to Alan Hale, director of the Southwest Institute for Space Research, "Rampant Scientific illiteracy in the general public is, in my opinion, one major cause of the current lack of opportunities for scientists. . . . A public that is ignorant of science, and of how science is done, is not going to support scientific research enthusiastically" (Greenberg 2001, 205). Alternatively, there are still a handful of technocrats, including the Italian economist, Giulio Sapelli, who agree with Henry Gantt's views and argue that a "council of sages" consisting of technical experts should govern (Bucchi 2010, 2–3).

There is some evidence that the absolute confidence of Americans and Europeans in science and scientists, measured by responses to the question, "Thinking about some different professions, how much do you think the following contribute to the well-being of our society?" declined between the 1970s and 2010, especially among American political conservatives. However, the ranking of scientists relative to other professionals remains stable near the top, with 93 percent of the U.S. population reporting in 2010 that scientists contribute a lot or some, compared to 95 percent for the highest rated group, members of the military, 77 percent for clergy, and 64 percent for business executives (NSF 2012, 7–29; Gauchat 2012, 175). International comparisons suggest that while Europeans vary from nation to nation, their confidence is slightly lower—that is, in Britain 88 percent responded positively in 2011(Ipsos MORI 2011, 19)—than that of Americans and their fears of the dangers of such technologies as nuclear power and genetically modified foods is significantly higher (Scientific American 2010; Bucchi 2010, 5–8). Asian populations have confidence in science levels similar to those of Americans, and they seem to be rising since the 1990s (NSF 2012, 7–29).

The evidence for rising obscurantism and opposition to scientific and technical progress thus seems to be very limited indeed, though it does seem true that especially in Europe, in spite of their high level of general support for science and technology, citizens are increasingly concerned about the safety of specific technologies, including nuclear power and genetically modified crops (Scientific American 2010; Bucchi 2010, 19–21). There does seem to be evidence that scientific literacy is weakly correlated with general positive attitudes toward science except among political conservatives, for whom a strong negative correlation exists (NSF 2012, 7–28, 7–29; Gauchat 2012, 169–170, 182). But there seems to be no correlation whatsoever between general scientific literacy and attitudes related to specific technical controversies (Allum et al. 2008, 36–37).

Two things do seem to be responsible for some of the decline in public confidence in scientific and technocratic expertise. First is the widely held perception that, "the independence of scientists is often put at risk by the interests of their funders" (Bucchi 2010, 23)—a perception that is supported by a wide range of studies (Oreskes and Conway, 2011; Greenberg, 2007; McAvoy 1999; Nelkin, 1975). Second is the closely related perception that scientists have become increasingly involved in producing proprietary knowledge that they themselves have a financial interest in promoting, rather than public knowledge in which they have no financial interest (Greenberg, 2007). Public unease about nuclear power probably stems from the first concern, while unease about genetically modified organisms (GMOs) is likely to be related to the second (Thackray, 1998).

While I would be hard-pressed to argue that greater scientific literacy would be a bad thing, I am deeply suspicious of the idea that a scientifically more knowledgeable citizenry and/or political elite would guarantee agreement regarding the solution to all or even many social and political problems. Nor do I believe that most scientists, technical experts, or political and social theorists in the West today share the nearly unbounded optimism regarding the ability of science to solve social problems that Taylor, Hoover, and much of several succeeding generations had.

Citizens and even some scientists may be skeptical about the objectivity of scientists and technocrats today, yet scientism and technocracy are probably more deeply imbedded in our industrial, business, public, and even private lives today than they were in the heyday of Progressivism. Sandra Harding was almost certainly right when she wrote in 1986, "We [in the West] are a scientific culture . . . scientific rationality has permeated not only the modes of thinking and acting of our public institutions but even the most intimate details of our private lives. . . . Neither God nor tradition is privileged with the same credibility as scientific rationality in modern cultures" (Harding 1986, 16).

In spite of, or perhaps even because of, their growing sophistication and modesty, and regardless of the public's degree of scientific literacy or their general attitudes toward science, scientists and technocrats and scientific/technocratic attitudes continue to play a major role in our lives. Moreover the tensions between expertise and democracy which manifested themselves so clearly in the careers of technocrats like Herbert Hoover and in the developing world during the 1960s through the 1990s continue to disturb both intellectuals and ordinary citizens. Debates over the roles of economic experts in the formulation of EU policies (Radaelli, 1999) or that of technical expertise in dealing with climate change (Oreskes and Conway, 2011) are just two obvious examples.

In this chapter I explore recent trends and experiments among technical educators and practitioners, some of which have been intended to mitigate negative characteristics and consequences of earlier technocratic attitudes. In the next I consider trends and experiments intended to integrate technical expertise more effectively into democratic politics. While I am hopeful about the growing sensitivity of technical expertise to an increasing range of cultural considerations and about the democratization of technical expertise, I have some concern that such trends might have the negative consequence of making it ever more difficult to question and/or resist some of the often unconscious core values of technocracy, including its tendency to measure human well-being or quality of life almost exclusively in connection with material considerations and notions of efficiency.

Changes in Engineering Education

ABET (the Accreditation Board for Engineering and Technology from 1980 to 2005) and its predecessor, the Engineer's Council for Professional Development (ECPD), have been accrediting academic engineering programs in the United States since 1936. ABET represents more than thirty specialized engineering societies, whose members are practitioners of those engineering professions, and since 1989 it has become the primary evaluator of engineering programs internationally. Because graduation from an ABET accredited program in the United States guarantees a graduate's ability to be licensed as a professional engineer in many states, and because graduating from an accredited program internationally guarantees that national recognition as a professional engineer can be carried to all nations that are signatories of the 1989 Washington Accord (at present, signatories include Australia, Chinese Taipei, Canada, Hong Kong, Ireland, Japan, Korea, Malaysia, New Zealand, Russia, Singapore, South Africa, Turkey, United Kingdom, and the United States, while China, India, and the Philippines hold provisional status and are in the process of meeting the requirements of the accord) there is a great incentive for engineering programs both within the United States and internationally to meet ABET accreditation criteria.

Initially the ECPD, and up to 2000, ABET accreditation criteria focused on providing courses that promoted technical knowledge in mathematics, the natural sciences, and engineering, and it specified nothing about the sixteen units of non-technical courses that it required starting in 1956 except that such courses "not include accounting, industrial management, finance, personnel administration or ROTC" (Prados 2007, 116, 241). But in its *Engineering Criteria 2000*, which was drafted in 1996, ABET changed its practices in two key ways. First, rather than emphasizing courses, ABET laid out a set of eleven areas of student learning outcomes (3a through 3k) to be assessed using a variety of instruments to be chosen by the departments seeking accreditation. Second, and most important for present purposes, among those areas ABET added the following areas to the prior emphasis on technical knowledge:

> 3c an ability to design a system, component or process to meet desired needs
> 3d an ability to function on multi-disciplinary teams
> 3f an understanding of professional and ethical responsibility
> 3g an ability to communicate effectively
> 3h the broad education necessary to understand the impact of engineering solutions in a global and societal context
> 3i a recognition of the need for, and an ability to engage in life-long learning
> 3j a knowledge of contemporary issues. (ABET 2000; Prados 2007, 255, 259)

In its most recent iteration in 2011, student outcome 3c has been modified to expand the requirements for knowledge beyond the purely technical as follows:

> 3c an ability to design a system component, or process to meet desired needs *within realistic constraints such as economic, environmental, social, political, ethical, health and safety,* manufacturability, *and sustainability*. (ABET 2011)

All of these criteria are open to interpretation—especially h, j, and the new version of c—and there is disagreement among engineering educators regarding whether teaching about several of them should be integrated into departmental courses or should be farmed out to humanists and social scientists. Moreover, there is significant fear among academically and politically conservative faculty and students that too much technical content must be sacrificed to respond effectively to the new demands within an undergraduate program, and that fear creates resistance. Nonetheless, the new criteria emerged out of the perceived needs of engineering practitioners. Moreover there is certainly evidence that the new criteria are responses to developments that were already under way within some engineering schools in the late twentieth century (Lynch 1997, 27–36) and that many engineering educators are taking the new criteria seriously (Herkert 2006; Felder and Brent 2003).

In part because it is less ambiguous than some and in part because there is a well-defined group (philosophers interested in ethics) outside of engineering who have an interest in promoting it, criterion 3f, which deals with engineering ethics has received a great amount of attention, extending even into graduate engineering education (Lucena et al., 2007; Newberry et al., 2011). Though the discipline of engineering ethics emerged in the 1970s in the United States, as late as 1984 engineering ethics courses were taught in only a handful of schools with engineering programs, and no engineering courses were identified as incorporating considerations of ethics (Lynch 1997, 28). By 1997, however, nine out of the top ten universities with engineering programs offered engineering ethics courses either within the engineering departments themselves, within a philosophy department, or within a Science, Technology, and Society (STS) Program. In a few institutions including Texas A&M, engineering ethics courses were required and in some, including Stanford and North Carolina, there was an STS requirement that could be satisfied with an engineering ethics course; but in most institutions engineering ethics remained an elective option (Herkert 2006).

Since the inception of *Engineering Criteria 2000*, there has been a growing trend toward the teaching of engineering ethics either in stand alone courses or in modules within engineering courses (Newberry et al. 2011). Wherever they are being taught such courses almost universally use a case studies approach (Lynch 1997, 33–34). In addition, interest in engineering ethics has led to the creation of a new journal, *Science and Engineering Ethics,* and to a new NSF Program, "Ethics Education in Science and Engineering," which focuses on graduate programs, new textbooks, and a vast array of on line resources (Herkert 2006; Newberry et al. 2011, 172–73). While it is hard to know yet how this new emphasis on ethics will impact technocratic attitudes, the response of graduate students to some of the NSF sponsored curricular experiments has been very positive.

Criteria 3h, 3j, and the newly formulated 3c are all substantially less clear than 3f, and there is considerably more disagreement about how to integrate them into the education of engineering students. Since the creation of Cornell's Science, Technology, and Society program in 1969, however, there has been a growing academic field whose function has been,

> to depict [science and technology] as complex socially embedded enterprises in which cultural, political, and economic values, as well as technical expertise, shape the directions of scientific research and technological innovation. In turn, of course, the products, and even the conduct, of science and technology affect cultural, political, and economic values and through them society and its institutions. (Cutcliffe 1990, 361–362)

In part because STS programs emerged out of the turbulent 1960s and 1970s, much of the early literature was critical of technology. STS programs were developed primarily at institutions with engineering programs, and the hope of many STS educators was "to bring humanistic perspectives [often hostile toward science and technology as it was being done] to the application and evaluation of technology" (in Cutcliff 1990, 361). Since the 1980s STS literature has become more balanced in its approach without becoming a mere cheering section for scientists and engineers, and engineering programs at places like Stanford, North Carolina, Lehigh and the University of Pennsylvania have begun to require that students take some STS courses. At other institutions, including MIT, Smith, and Harvey Mudd College there have been institutional or departmental initiatives to bring STS content into core science and/or engineering courses.

At Harvey Mudd, for example, for classes graduating between 2003 and 2012, all students were required to take an "Integrative Experience (I.E.)" course which combined technical with humanistic or social scientific content in order to graduate. Some of those courses, including one on Alternative Energy Sources were co-taught by technical and non-technical faculty; but every technical department was required to provide at least one course on its own. Engineering provided an environmental engineering and a transportation engineering course that satisfied the requirement. When a curriculum revision was done in 2008 to be initiated with the class of 2013, the Integrative Experience requirement was dropped and replaced by a faculty imposed requirement that all technical core courses incorporate humanistic or social scientific elements. That new requirement is being implemented only irregularly so far, but many of the I.E. courses survive without substantial drops in enrollment.

Though not specifically developed to meet criteria 3h, Engineers Without Borders—USA (EWB–USA) was created by Civil Engineering Professor, Bernard Amadei, at the University of Colorado at Boulder after he and a group of his students had responded to a request from San Pablo, Belize, to help design and implement an improved water supply in 2000. EWB–USA was chartered in 2002, and in 2013 it had nearly 300 chapters in as many U.S. engineering programs as well as projects in forty-seven different countries. Each project is a response to a request from a community in the developing world that incorporates a substantial engineering design component. The project is led by a professionally qualified mentor and is implemented by a team of engineering students. It demands an assessment of the community's resources and constraints. It demands partnering with a local organization—usually an NGO. And the community must demonstrate that it has the financial, administrative, technical, labor, and material capacities to operate and maintain the constructed facilities on a long-term basis without outside assistance (though the EWB–USA may provide some of those capacities through educating members of the local community). EWB projects automatically respond to 3h and modified 3c criteria, they are involving substantial and increasing numbers of engineering students, and the skills are being directly applied to practice.

Beginning at Purdue in 2004, Engineering Education graduate programs have begun to emerge both in the United States and in Europe. Though most engineering educators maintain an "objective" and technocratic approach, some practitioners, such as Gary Downey at Virginia Tech and Brent Jesiek and Alice Pawley at Purdue bring both engineering and STS perspectives to their work and are preparing course materials that integrate STS concerns into technical courses or which create courses in the liberal arts that speak directly to engineering issues (i.e., Lucena 2013, Lucena 2010). Student responses to one such course, *Engineering and Sustainable Community Development*, organized by Juan Lucena, Jen Schneider, and Jon Leydens at the Colorado School of Mines, suggests that they have the capacity to transform engineering students' attitudes, making them vastly more sensitive to the non-technical elements associated with development projects in particular and reducing their confidence that there can be purely technical solutions to social problems (Lucena et al. 2010, 185–202).

Similarly, spurred on by a number of factors, including an awareness that women and minority students are retained in programs that stress social relevance and social justice, a few engineering educators and engineering programs have been incorporating issues of social justice into engineering curricula (Lucena 2013, 8–9). Though "social justice" is a term about which there can be much disagreement, Juan Lucena offers a definition which is directed at engineers:

"Social justice practices, including those by engineers, should attempt to an equal distribution of rights, opportunities and resources in order to enhance human capabilities and reduce the risk and harms among the citizens of a society" (Lucena 2013, 10). Organizations, including the Engineering for Social Justice Network (ESJN) established by the Chemical Engineer and Engineering educator, Caroline Baillie, in 2004 and the Liberal Education/ Engineering and Society division of the American Society for Engineering Education, have provided support and encouragement to those who are developing courses and modules on social justice to incorporate into technical engineering courses (Riley 2013, 58–59). Moreover, in 2012 a new journal, the *International Journal of Engineering, Social Justice, and Peace* was launched (Nieusma 2013, 35).

Incorporating issues of social justice into engineering education faces even greater resistence among traditionally trained engineers than incorporating other non-technical issues because engineers still tend to identify themselves and their work as objective and politically neutral, and they view social justice, unlike such topics as the economic and social contexts of engineering projects, as an inherently subjective and political domain (Cech 2013, 67–84). What makes the situation even more problematic is that ESJ advocates explicitly deny the "a-political myth" regarding engineering (Leydens 2013, 188–190; Sakellariou 2013, 253–254). In spite of this resistence, however, social justice is making its way onto the agendas of even such traditional engineering strongholds as the National Academy of Engineering (National Academy of Engineering, 2008) and into curricular experiments at a few schools including the Colorado School of Mines, Olin College, and Smith College, with an expectation that it will continue to spread as the foci of engineering ethics and the broader context of engineering have. As a consequence, one expects or at least hopes that even the traditional lack of interest among engineering-trained technocrats in distributive justice will diminish over time.

The Evolving Character of Economics Education: A Narrowing of Focus

Several circumstances have produced trends in economics education that have narrowed rather than broadened the education of economists throughout the twentieth century, though there is some indication of developments within economics research that have the potential to reverse those trends. The content of engineering education has always been responsive to the requirements of professional engineers outside of educational institutions, and engineering has had an organization, ABET, capable of enforcing external demands on engineering curricula. Economics, on the other hand, developed within an academic environment which has been focused on academic freedom and responsive to a long-term trend toward increasing specialization. Even today, although the American Economics Association created a Committee on Economics Curricula in 1976, that committee has not even hinted at a desire to enforce any curricular demands on economics departments.

It was only at the beginning of the twentieth century and under the influence of Alfred Marshall, who pushed economics toward greater mathematical rigor, that economics was established as a special discipline of its own in place of the more comprehensive subject of political economy. At Cambridge, Marshall created a new tripos for economics in 1903. Until that time, economics was examined under the Historical and Moral Sciences Tripos. In the United States, economics replaced political economy early in the twentieth century, but in many colleges economics courses continued to be taught within broader departments of social sciences. At Pomona College, whose catalogs have been explored by my student Xaiolong Wang, for example, economics courses were taught within a department of economics and sociology until 1929. Even when more specialized departments of economics emerged, prior to WWII, the orientation of the vast majority of economics courses was that of what is now called institutional economics, which paid special attention to the social and institutional contexts in which economic decisions are made and which tended to use verbal rather than mathematical arguments.

During the 1930s the mathematical modeling of economic systems became an important part of research in economics, but it was only during the 1940s that mathematics, especially statistics and modeling, moved into the core of economics education, augmenting the institutional focus of American economics and buttressing claims regarding the scientific character and rigor of economics. At Pomona College, for example, the first statistics class entered the economics department in 1942 and the first course on economic research entered in 1947. What had previously been subfields within economics, including economic history, history of economic thought, industrial relations, and business management, gradually became established as their own departments or moved into history departments (Morgan 2003, 278) and disappeared from economics departments (Aslanbeigui & Naples 1997; Foggarty and Naples 1998). More than before, economic claims were quantitative, seemingly universal, and grounded in rationally calculated self-interest—whether the self being considered was an individual person or firm (the subjects of microeconomics) or the national economy (the subject of macroeconomics). Within the new economic emphases, laborers, for example, who might have been seen primarily as people with special needs, interests, and demands in the industrial relations context, became forms of human capital to be dealt with in precisely the same way as any other form of capital.

A second trend began to show up in American economics departments after the end of WWII. From its emergence out of political economy in the early twentieth century, most economic ideas and theories had been concerned with applied issues connected to public policy or business management; but as long as economics was not highly mathematical, economists had difficulty in claiming any special expertise, especially in the public policy domain. With the growing mathematical sophistication of the field, not only did economics students have to take more mathematics, but the mathematical elements in economics branched into two different trajectories. A few economists turned toward the creation of pure and comprehensive mathematical economic theory which became highly abstract and relatively divorced from direct empirical testing. The majority, however, continued to be more interested in consulting and policy work. They tended to develop mathematical models that were narrower in scope but more directly related to empirical data. The work of the latter group was very useful, and economists began to replace engineers as the dominant voices of technocracy during the 1960s , in part because their modeling skills could be applied to virtually any kind of problem, as MacNamara, Hitch, Rostow et al. had demonstrated during and after WWII. Their work was usually quite divorced from pure theory, although economic modeling gained authority at least in part because of the scientific reputations of the pure theorists.

There were multiple schools of economics during the first half of the twentieth century, depending on how economists understood the notion of "utility," which was responsible for generating consumer demand. Virtually all Western schools, however, were grounded in Adam Smith's assumptions that under normal conditions, free markets allocate economic resources effectively through some kind of "invisible hand" and that in growing economies, labor demand will ensure that wages are sufficiently above those required for mere subsistence so that any increase in productivity will be divided between labor and capital, guaranteeing that the standard of living will rise for all members of society without the necessity of worrying much about distributive justice. Such assumptions, as we have seen, were at the foundation of early Scientific Management's lack of concern for the distribution of the goods generated by increased efficiencies in production as well as among the basic assumptions of the theory of economic development promoted by Walter Rostow and implemented in most developing nations except for the "Asian Dragons," where opposition to economic concentration helped to create strong internal demand which fueled sustainable economic growth.

Under extremely limited conditions one can demonstrate that a two party "free market exchange" cannot diminish the utility of either party while it must increase the utility of at least one party. In a sense, then, free market exchanges can be shown to increase net utility, but not to maximize it; and that can only be demonstrated when the following improbable conditions pertain: 1) No one but the two parties directly entering into the exchange are affected (i.e., there are no "externalities,") 2) both parties are perfectly informed regarding the consequences of the exchange, 3) both parties are completely and equally free to make or not make the exchange. While condition 3 may occasionally be met, condition 1 is almost never met, and condition 2 is seldom met.

It is also true that growing economies may usually have generated substantial labor demand in Adam Smith's day; but during much of the nineteenth century and virtually all of the twentieth century, economic growth has depended primarily on increases in productivity generated by technological innovation. As a consequence, economic growth has often been associated with declining labor demand; so wages have frequently declined during periods of economic growth. In the United States, for example, between January of 1973 and January of 1995, the economy grew by about 3 percent per annum while average hourly wage rates declined 14.5 percent from $8.63 to $7.38 in constant 1982 dollars (Bureau of Labor Statistics data, 1998). As a consequence, the standard of living for hourly wage workers in the United States declined. Thus classical economic theory, as interpreted by advocates of Scientific Management and of development theory in the mid-twentieth century, failed to understand the problems of income concentration attendant on economic growth through technological innovation.

For a very brief period between 1938 and 1947, Keynesian economics, which granted a much more extensive role to government expenditures as a way of stimulating demand, was taught and sometimes emphasized in U.S. introductory economics courses. But the similarities between Keynesian arguments and socialist arguments bothered many Americans. By 1947, rampant anti-communism led conservatives to condemn the few principles textbooks that advocated Keynesian ideas (Colander and Landreth, 2007) and it effectively disappeared from undergraduate American economics curricula along with references to Marxism for about a decade. Except at a handful of institutions, including the University of Massachusetts at Amherst and the University of California at Riverside, where radical economists were in control, courses on comparative economic systems were rare. Even at a small liberal arts college, such as Pomona, comparative systems courses disappeared in the 1950s at the height of American anti-communism, to reappear later in the 1960s.

Especially as a result of anti-communist intimidation, the pluralism of approaches to economics which had been a feature of American economic education during the inter-war years gave way in the vast majority of American institutions to a dominant tradition of neoclassical economics. According to Mary S. Morgan,

> This approach continued to gain credibility through the first half of the twentieth century, as the characteristics of what was to become the full-fledged neoclassical economics of the third quarter of the century—namely, formal treatments of rational, or optimizing, economic agents joined together in an abstractly conceived free market, general equilibrium world—were worked out. This abstract account became widely adopted to the exclusion of other approaches, however, only during the second half of the twentieth century. (Morgan 2003)

In spite of what has seemed to many outside of economics, and more recently to a few within economics, as a completely unrealistic view of individuals as nothing more than rational calculators and optimizers of self-interest, this neoclassical paradigm has been all but hegemonic from the late 1970s to the early 2000s in the United States (Morgan 2003, 295–298). Its practitioners have viewed themselves as objective and "neutral" in spite of strong ideological roots. Moreover they have viewed themselves as working within a dominant scientific paradigm and have therefore viewed alternative approaches as "unscientific" by definition. In addition, they have seen their blocking of alternative views from mainstream journals as completely justified (Fogarty and Naples 1998, 233).

While economics departments now typically offer a wide range of topical courses, including labor economics, developmental economics, finance, public finance, even occasionally "religion from an economic perspective," since the 1970s, neoclassicism has provided the analytic tools for studying virtually all of these topics at most American institutions. It is for this reason that I would argue that economics education has become narrower rather than broader during the last several decades as engineering education has become broader.

Only in the last few years have behavioral economists, evolutionary economists, experimental economists and neuroeconomists begun to report research results which challenge the core assumptions of the neoclassical school (Zak, 2008). Until their work finds its way into undergraduate economics curricula, we can expect economics education to remain extremely narrow and ideologically biased.

A Comment on STEM Education in General

Since technocratic attitudes are often expressed by mathematicians and scientists outside of engineering and economics, some attention should be paid to the S and M (Science and Mathematics) elements of contemporary STEM Education. Very few uniformities can be found within curricular changes in the sciences, mathematics, and computer science in recent years except for a tendency to try to pack more technical content into undergraduate majors. General education requirements in colleges and universities continue to ensure some breadth of education for science and mathematics students; but a clever and determined student at many liberal arts colleges can negotiate the system to take four or fewer courses outside of STEM departments. Moreover it is a sign of narrowing focus when Maria Klawe, president of Harvey Mudd College, one of the premier colleges of science, mathematics, computer science, and engineering in the United States, makes a point of arguing that "From my perspective, computer science is something that everyone needs. [Coding] is way more important than American History." (recode.net/2014/05/28/how-better-tech-education-can-unlock-a-half-trillion-dollar-opportunity) While there may be contexts in which this statement is true, it certainly has the tendency to lessen the value of non-technical knowledge and to promote technocratic attitudes.

The Evolving Character of Think Tank Personnel and Advice

When RAND came into existence as a separate not-for-profit corporation in 1948, it had a single client, the U.S. Air Force, and for its first four years it advised only on defense-related issues though its initial charter declared its broad purpose to be "To further and promote scientific, educational, and charitable purposes, all for the public welfare and security of the United States of America." In late 1948 its staff of approximately two hundred members was made up almost exclusively of mathematicians, physicists, engineers, aeronautic scientists, chemists, and economists, with a few statistically oriented psychologists. In 1952, a Ford Foundation grant freed a few staff members to undertake a few small non-defense-related projects.

Sixty years later, RAND had over 17,000 staff members in over fifty countries. Moreover, it currently advises U.S. government agencies, agencies of other states, and private clients on policy issues in connection with twelve different broad areas: Children and Families, Education and the Arts, Energy and Environment, Health and Health Care, Infrastructure and Transportation, International Affairs, Law and Business, National Security, Population and Aging, Public Safety, Science and Technology, Terrorism and Homeland Security. In each of these fields it claims to provide studies which are "objective, independent, and balanced" and which present alternative policy perspectives "fairly and accurately" while "weighting them according to merit" (RAND 2009, 12).

A sampling of staff members suggests how RAND's corporate culture has changed since the early 1960s largely, though not exclusively, in response to an expanded range of clients and areas in which policy advice is offered. I used RANDs web sight in 2013, which provided an alphabetical list of staff with short biographies. Nine hundred and thirty-five staff members were listed in this directory. Each entry offered a one paragraph biographical statement, an education entry which normally included the disciplines of bachelors, masters, and most advanced degrees, a RAND supplied list of areas of research, and often a list of recent publications. For each entry, my research assistant, Sean Butler, and I identified the following, when explicit or easily inferred: gender; national origin; educational background (bachelors, masters, and terminal degrees); primary and secondary areas of research (using RAND's self identified twelve policy areas plus a category for methodology when that appeared to be the primary emphasis); and geographical area of special expertise, when stated. In that year, nearly 29 percent of all RAND research was in the area of health and health care, approximately 17 percent dealt with national security, terrorism, or homeland security, 15 percent dealt with non-defense-related science and technology, 13 percent with law and business, 8 percent with international affairs, 7 percent each with education and infrastructure, and less than 5 percent each with children and families, energy and environment, population and aging, or public safety.

If it were the case that RAND personnel only worked in a single field related to their educational backgrounds then each of the twelve areas of investigation claimed by RAND might have its own largely independent culture; and defense-related studies might, for example, be as narrowly construed as they were in the early 1960s. But such is not the case for several reasons. First, a whopping 80 percent of the RAND staff work in two or more of the twelve broad research areas, in principle, bringing the attitudes, ideas, and interests associated with one domain into the others in which they work. In that connection, it is important that RAND has, at least since the late 1970s, been a pioneer in interdisciplinary and what has come to be called "transdisciplinary" research practice—that is, practice in which "scholars and practitioners from [multiple] academic disciplines and non-academic fields work jointly to develop and use novel conceptual and methodological approaches that synthesize and extend discipline-specific perspectives, theories, methods, and translational strategies to yield innovative solutions to particular scientific and societal problems" (Stokals, et al. 2013, 6; Holmes et al., 2008, S182—S192). In fact, RAND calls out its emphasis on multidisciplinary, synthetic analyses as one of the features which define its unique institutional legacy in its *Standards for High-Quality Research and Analysis* (RAND 2009, 15–16). Such practices have subsequently spread to virtually all major policy research organizations today.

The fact that about 14 percent of RAND staff members are graduates of or graduate students at the Pardee RAND Graduate School, which emphasizes multidisciplinary research, reinforces the emphasis on transdisciplinarity brought to each subject of research. In addition, RAND now has a substantial number of geographical and national area specialists to ensure that local cultural and political situations are considered when relevant. In 2013 those included approximately one hundred Asian specialists, fifty middle Eastern specialists, fifty Latin American specialists, and thirty African specialists. Thus many of the shortcomings of studies which expressed the early and more narrowly technocratic emphases at RAND have been eliminated in more recent studies done by RAND and almost all policy advising organizations.

But there is an attendant danger. In spite of the breadth of disciplines that may be represented, policy analyses still focus primarily on issues that can be quantified and they virtually all focus on delivering material goods or services efficiently and at low cost, rather than on such issues as provider satisfaction, or fairness. If one considers the disciplines of highest degree among RAND employees and compares them to the dominant methods used, for example, one finds that those who focus on economic analysis are more than double the number with their highest degrees in economics. Similarly, though less than 10 percent of RAND employees have degrees in mathematics, statistics, or operations research, nearly 20 percent are engaged in developing mathematical models and applying game theory to topics which are not primarily economic. In addition, virtually every discipline with the possible exceptions of engineering specialties and the exact sciences contains contending sub-disciplines openly based on preferred methodologies and divergent social and political preferences. Neoclassical economists trained in Ivy League universities, Chicago, and Stanford, for example, are unlikely to see as objective and balanced those studies carried out by Marxist trained political economists from The National Autonomous University of Mexico and *vice versa*. Indeed there has been a distressing history of intimidation among orthodox economists in many American economics departments to ensure that students have little or no exposure to heterodox views (Colander and Landreth 2007; Tharp 1970).

Psychologists and educational policy analysts whose primary interests are in measurement and statistics are often more likely to share viewpoints with mathematicians than with therapists or teachers. Similarly, someone trained in a tradition of deontological ethics is unlikely to see as objective and balanced studies carried out by those committed to a utilitarian or naturalistic ethical tradition and *vice versa*. The fact that a number of disciplines are represented on a study team does not therefore guarantee that the viewpoints present are significantly divergent. It is possible that the claim that multiple perspectives inform some organization's studies may lend legitimacy and make the resulting policy implications more difficult for clients and the public to resist while it does not produce either the kind of balance or the breadth of views implied by the organization's claims.

Chapter Seven

Continuing Concerns about the Role of Technical Expertise in a Democracy

From the early days of Scientific Management to the present, there has been disagreement and concern regarding the relationship between technical expertise and democratic governance (Fischer, 1990; Brown 2009). A dominant position among early public administration advocates, including Woodrow Wilson and Frank Goodnow, was that expert implementation of policies could be totally divorced from the political establishment of policies. From this perspective there is no relationship, either positive or negative, between the democratic setting of policy and administrative expertise. In principle, this remains the guiding assumption for the city manager—city council form of government in which an elected council presumably makes policy and the appointed city manager determines the most effective way to implement it.

In practice, however, this form of government operates more like Morris Cooke and most early members of the New York Bureau of Municipal Research envisioned. City managers and various commissions appointed by the Council, at least in part for the members' expertise, provide information and make recommendations regarding policy options to the Council, which represents the citizens. Using that information and in the light of the experts' recommendations the council then determines what policies to adopt, and the manager and commissions implement those policies, whether or not they were policies favored by the experts. When exceptions to the policies are requested, they must usually be approved by the elected council, again based on recommendations made by the experts. At the local level, this relationship between expertise and democracy satisfies most citizens, but because managers and commission members serve at the pleasure of elected councils there is a constant likelihood that the expert advice will be influenced by the political orientation of the council and will not have the independent, objective, character which technocrats claim to provide.

At the state and federal level and beyond, this process is usually complicated in at least three ways. First, advice is often sought from appointed adhoc groups or, in recent years, from professional policy research organizations usually identified as think tanks, or from quasi-governmental organizations such as state bar associations or medical associations, which are neither appointed nor elected by any government body but which may adopt and enforce policies which carry governmental authority. Second, in many cases the elective legislative, or law and policy-making body is, in principle, separate from an elected executive branch of government which presumably implements the laws and policies. But the executive branches of governments usually have greater access to advisory expertise; so they have, in effect, a very large role in promoting as well as in enforcing legislative policies. Furthermore the boundary between policy formulation and implementation is often very blurred. In the United States, for example, a great deal of policy is de facto established by executive order or by appointive bodies. Moreover, there are many domains, including monetary policy, in which experts—in this case, members of the Federal Reserve Board—are granted authority by the executive and legislative branches of government to both determine and implement policies which may have a major impact on society.

Because scientific and technical advisory bodies usually serve at the pleasure of political actors, the likelihood that they will at least to some degree reflect electoral politics rather than be independent of political attitudes remains significant. Alternatively, if they do not reflect the views of the current political regime, the policy suggestions may be ignored or distorted by those in power. Both of these possibilities were at the bottom of the severe antagonism between large segments of the scientific community and the George W. Bush administration. In 2004, scientists from the Union of Concerned Scientists and a large number of members of the National Academy of Sciences charged that the administration, "manipulated scientific advisory committees, altered and suppressed reports by government scientists, and misrepresented scientific knowledge in contentious areas such as global warming, air pollution, and reproductive health" (Hayden 2005; Brown 2009, 1–5).

From a participatory democratic perspective the fact that policies are often based on information and advice provided by technical experts may allow technocrats to formulate all practically available policy options. So even when policies must be formally adopted or otherwise authorized by elective representative bodies, this situation raises the possibility suggested by Eisenhower regarding national security—that is, that policies may become the captive of a technocratic elite. At the extreme end of the technocratic spectrum, represented by Henry Gantt among early Scientific Management experts or more recently by the former Italian Minister of Health, Umberto Veronesi, of course, the capture of public policies by technical experts is something to be hoped for rather than avoided.

In what follows, I will assume that neither the total divorce between democratic policy setting and expert implementation of policies nor governance by a technocratic elite is a realistic possibility. Instead I will consider a variety of ways in which advocates of both representative and participatory democracy and of the use of technical advice propose to find appropriate and sustainable relationships between expertise and democratic governance, simultaneously taking advantage of technical expertise and containing its influence.

The Confrontational Pattern of Relating Expertise to Democratic Politics

It is often the case that the bounded rationality within which technocratic policy advice is generated and offered is inconsistent with features of democratic political life. We saw in chapter 4, for instance, that early expert studies regarding the closure of military bases failed to consider the perceived needs of members of congress to please their constituents (p. 91). Within the bounded rationality of policy advisors, such a perceived need seemed irrational and therefore ignorable, while within the bounded rationality of the elected official, acting in a manner that would promote reelection was completely rational. In this case it made sense for political actors to ignore or act contrary to the expert advice which they received even though from the policy expert's point of view such actions worked against the general public interest.

One of the most interesting situations occurs when expert policy advice focuses on the general public interest which may be in obvious conflict with the interests of smaller groups of citizens, who may mobilize to oppose the policies promoted by the experts. Among the most common and annoying of such situations are those associated with what has come to be called the NIMBY—Not In My Back Yard—syndrome in which local interests oppose the siting of some facility which may have some negative impact on their community but which is intended to serve the greater public good. With rare exceptions both experts and representatives of the general public have viewed the Nimby syndrome in a completely negative way. According to the Southern California Waste Forum, a group of public officials and waste management experts,

> The Nimby syndrome is a public health problem of the first order. It is a recurring mental illness which continues to infect the public. Organizations

> which intensify this illness are like the viruses and bacteria which have, over the centuries, caused epidemics such as the plague. (McAvoy 1999,1)

Yet Nimby opposition to expert-devised policies is completely consistent with the rational interests of local groups; and in many cases it has improved the quality of decisions even from the perspective of the general public by raising issues and by generating policy options that expert advisors had not considered.

The goal of technocratic policy making is to begin from a comprehensive understanding of the problem to be solved, but as Robert Dahl has argued, "Precisely because the knowledge of the policy elites is specialized, their expert knowledge ordinarily provides too narrow a base for the instrumental judgements that an intelligent policy would require" (In McAvoy 1999, 9). So confrontation and opposition is sometimes a useful strategy for allowing "citizen rationality" a role in improving policies recommended through the relatively narrow views of experts.

Detailed case studies of the siting of nuclear power plants in New York as well as in France and in Germany by Nelkin (1971; 1981), of the construction of a new runway at Logan International Airport in Boston by Nelkin (1974; 1975), and of the siting of hazardous waste processing facilities in West Virginia by Hunter and Leyden (1995) and in Minnesota by McAvoy (1999) have clearly demonstrated that it is not generally true that Nimbys are "overly emotional, uninformed, and unscientific in their opposition to these facilities, [and] motivated by narrow, selfish interests" (McAvoy 1999, 3). In the case of local opposition to the siting of a nuclear power plant on Cayuga Lake in upstate New York, for example, leadership came from a group of seventeen scientists at Cornell University who raised issues of thermal pollution and radiation hazards that had not been adequately considered by the experts of the New York State Electric and Gas Company (Nelkin 1971, 248–53).

In the Logan Airport expansion case, initial opposition came from the working-class Italian neighborhood that would have been most impacted by the noise pollution from the runway, but they were soon joined by the City of Boston, which provided expertise that countered the Airport Authority's consultant's reports which insisted that the new runway was not only needed but that failure to build it "would relegate Boston to a 'second rate' airport and have 'disastrous effects' on service to sixty five percent of the 267 cities served by Boston" (Nelkin 1975, 46). The city's experts challenged the airport's consultant's estimates of increased demand, which had simply ignored their own recent data which showed a downturn in passenger demand, and they demonstrated that even if there should be a major increase in demand, it could easily be met by rescheduling commercial flights and by redirecting some general aviation flights (private, small plane, flights) to smaller local airfields. In this case it became extremely clear that expert advice, far from being impartial and objective, is frequently shaped to legitimize the interests of whatever client the experts are being paid by. Even supposing policy analyses to be unbiased, "the extent to which technical advice is accepted depends less on its own validity and the competence of the expert, than on the extent to which it reinforces existing positions" (Nelkin 1975, 52–53). Moreover, the consideration of a change in general aviation by the city's consultants demonstrated the ability of alternative perspectives to generate new policy options that could be acceptable to both the initial policy advocates and their opponents.

In the Minnesota waste treatment case, the state government, following the advice of the Minnesota Waste Management Board (WMB) and the Legislative Commission on Waste Management, proposed to establish a hazardous waste treatment facility in a rural part of the state after a new federal definition of hazardous waste in 1978 made it impossible for a major business, Minnesota Mining and Manufacturing (MMM), to continue to dispose of its waste in a landfill near St. Paul. MMM then had to resort to the expensive alternative of shipping its waste to out-of-state sites which charged much more to treat or bury the waste. Directed to consider all feasible and prudent ways of dealing with the problem of hazardous waste, the WMB produced a *Draft Certificate of Need* that included, in addition to special landfills and treatment facilities that would recover heavy metals and other substances for future use, bans on the production of certain wastes, and restrictions on businesses that would reduce other hazardous wastes. MMM played a major role in opposing any restrictions on business, arguing that they "would further discourage new industry from locating in Minnesota, and would be another reason for existing industry to reduce or discontinue operations in Minnesota" (In McAvoy 1999, 33).

Given their awareness of the importance of business support to the legislators who would be enacting any policies they recommended, the WMB eliminated from consideration all proposals that would restrict businesses. Thus, even before any proposal was forwarded to the Legislative Commission, business interests played a significant role in shaping the range of possible policy options considered by the experts on the WMB. Elite environmental groups like the Sierra Club also had clout with the legislature and WMB. In return for technical requirements regarding impacts on ground water and protection of local fauna and flora, these environmental groups agreed to support the WMB proposals by countering local Nimby opposition. So a process that was intended to serve the general good, was in some ways hijacked by business and other elite interests at its very inception and prior to the formal recommendation stage. Even more openly and explicitly than in the Airport expansion case above, expert advice, while nominally intended to serve the general interests, was far from being objective and balanced. Rather, it was directed to serve special client interests.

The act which created the WMB required that the WMB hold public hearings on their proposals. At the end of those hearings the WMB conducted a survey of those who attended which was given only internal circulation. Of members of the public who attended, 85 percent answered yes to the following question: "Should the state place a higher priority on hazardous waste reduction than on other methods of waste management, such as waste treatment or proper disposal, by concentrating its activities more on encouraging hazardous waste reduction than in developing hazardous waste facilities?" Similarly, 67 percent answered no to the following question: "Should the state develop and operate its own 'integrated' hazardous waste facility, including disposal, treatment, and temporary waste storage?" (McAvoy 1999, 35) In spite of the obvious public preference for waste reduction and opposition to a state run disposal facility, the WMB was not obligated to respond to what it learned in the public hearings that it was forced to hold. It thus spent almost no resources on developing waste reduction strategies and it ended up proposing a state run integrated hazardous waste facility in a rural county where the cost of acquiring land by eminent domain would be relatively low, reducing the costs to waste producers.

Responding to citizen complaints, the Minnesota legislature voted in 1984 to place a moratorium on the process and directed the WMB to do a more detailed inventory of the types and amounts of waste being produced and to consider more alternatives to landfill disposal and the costs of those alternatives. In 1986 the WMB came back with a new plan. While admitting that the amount of hazardous waste being produced was not presently enough to justify building an in-state facility, they projected increases that would justify a facility by the time that it could reasonably be sited and constructed. Moreover, now they proposed a stabilization and containment facility in which waste would be mixed with concrete and stored above ground where it could be easily monitored for leaks. The cost of such a facility to meet projected needs for a period of twenty years would be low enough to make it a competitively priced option for businesses in the state on the assumption that anticipated new EPA regulations would soon increase the price of alternatives.

Businesses like MMM and Honeywell would have preferred the cheaper landfill option and opposed the new facility on the grounds that their costs for using the facility would be higher than those for shipping their wastes out of state. In fact, MMM threatened to boycott the new facility, and, since it produced nearly a third of the toxic waste in the state, without MMM use the facility could not initially be self-supporting. The new plan did, however, appear to respond to the Sierra Club and other elite environmental groups. According to the Chair of the WMB, "Because of the high standards Minnesota has set for this facility, it will be more protective of the environment than any existing or planned waste treatment facility anywhere" (McAvoy 1999, 44).

In order to head off Nimby opposition, the WMB proposed a voluntary siting plan in which county boards of supervisors could volunteer their county for a site in return for a $4000/month payment, and two rural counties, Red Lake, and Koochiching, far from the location of waste production in Minneapolis and St. Paul, responded. The financial incentive did give rise to local groups which supported the WMB proposal; but it did not eliminate opposition. Red Lake county actually used some of its funds from the WMB to hire an independent engineering firm to evaluate the local impact and study any potential geological problems. In Koochiching county, opponents funded their efforts with bake sales while the WMB spent more than $100,000 on public relations to sell its point of view (McAvoy 1999, 77). In fact, it seems fairly clear that the WMB's public relations efforts backfired, creating distrust. Eventually, opposition groups managed to scuttle the proposed facility entirely.

At least a major cause for concern among opposition groups had to do with their perception of risks associated with the transportation of hazardous wastes the roughly 250 miles from its sources in the Minneapolis-St Paul area to the site of the facility. While the experts' "objectivist" assessment of risk was based on the probability of an accident, given the frequency of accidents per mile traveled by commercial carriers and the probable costs associated with attendant damage and cleanup, local citizens' "constructivist" assessment of risk was reached in conversations and almost inevitably focused on psychological and intangible issues that included the level of comfort that citizens have with any objective risk for a project whose value they questioned in the first place (Bucchi 2006, 78–79). Some even opposed the siting for aesthetic reasons. According to one elementary school teacher, "I chose to live up here for the clean water and the clean air, and I guess that is what we have going for us in Northern Minnesota. We have the environment, the health, all these things going for us" (McAvoy 1999, 94). The presence of a waste treatment facility in their midst seemed to strike at their very reasons for locating in the region. While such considerations may be a-rational, they are real in the minds of citizens, nonetheless. In spite of their legitimacy, such considerations are not usually effective in policy debates; so policy opponents must usually find alternative ways to justify their positions.

As in the Airport expansion case, one of the most valuable services of Nimby opposition to the siting of a stabilization and containment facility was to both challenge the assumption of need for the facility and to focus attention on policy options that had not been imagined or adequately explored. Ultimately, local opposition groups both demonstrated that the proposed facility could not be economically viable without extensive long-term state subsidies and they promoted the search for waste reduction strategies that led to the State's passage of the Toxic Pollution Prevention Act in 1990. This act, which provides financial incentives and support as well as technical assistance for businesses to reduce or eliminate wastes, was successful in lowering toxic wastes by 28.7 percent between 1992 and 1996 at the same time that the Minnesota economy was growing by about 5 percent per annum (McAvoy 1999, 128–131).

One downside of the confrontational style of relating democratic considerations to expert policy advice is that it is inevitably reactive. Citizen participation in policy considerations is only engaged when community groups feel seriously threatened in some way by a proposed policy; so citizens are virtually never capable of initiating policy discussions. A second downside is that it usually depends on the capacity of communities to generate voluntary, grass roots organizations, which is not always easy. A third downside is that effective opposition often depends on the presence, identification, and availability of alternative expertise. That is, Nimby opposition to a policy can easily be dismissed as nothing more than visceral and self-serving obstructionism unless opposition can be justified by an argument that the greater public good could be better served by an alternative identified by the opponents to the original policy. Moreover, the criteria for evaluating the alternative must be understood as "rational" rather than simply "emotional."

Methods of Linking Expertise to Democratic Politics (1): The Structural Approach of the European Union

The European Union (EU), which came into existence under that name in 1993, has given rise to intense debates regarding the role of technical expertise in representative democratic politics, with some authors arguing that under the Union, Europe is experiencing a "Democratic Deficit," while others argue that the structure of the EU is such that processes which are seemingly independent of ordinary democratic control are sufficiently accountable to elective bodies (Moravcsik 2004; Gottwald 2005). The tension between technocratic tendencies and democratic ones in the EU is particularly visible for at least three reasons. First, the traditions of democratic politics—whether direct or representative—vary tremendously among the twenty-eight member states, leading to inevitable disagreements about how much power can legitimately be delegated to institutions that are largely insulated from political control (Majone 1998, 5–28). The issue of inconsistent standards is complicated by the desire of all member states to hold on to core elements of national sovereignty; so there is an insistence that on many issues, decisions of the European Parliament be subject to veto by the Parliaments of the member states, and that position has been confirmed by a vote of the people in most member states.

Related to the issue of appropriate standards but more important for present purposes is the pre-EU background of European cooperation, starting with the creation of the European Coal and Steel Community (ECSC) in 1952 and the European Economic Community(EEC) and European Atomic Energy Community (Euratom) in 1957. While the overarching goals of these communities were established by politically negotiated treaties, the detailed policies of each were created and implemented by appointive Commissions whose members were chosen for their expertise. The choice to join was made by election in each new member state, but once a member, the nation was obliged to accept the commission's actions; so there was relatively little democratic oversight of commission actions, which could generally be characterized as technocratic—that is, based on expertise.

During the early 1960s France began to express concern about the threat to its national autonomy by the supranational powers invested in the three commissions mentioned above; so in 1967 a Merger Treaty created a single *European Community* (EC) with a single Commission to replace the three earlier ones. The new Commission had one appointive member from each participating nation and it acted much as an Executive Cabinet, with each member taking responsibility for some policy area and leading the efforts of a committee of experts (Directorate General) responsible for forwarding policy suggestions to the Commission and for implementing policies adopted by the Commission.

Though the Commission as a whole involved representation of the various national interests and could be seen as in some sense indirectly responsible to democratically elected officials, its mode of operation continued and still continues as de-facto technocratic. There have been complaints that, "EU technocrats working in splendid isolation tend to produce too much regulation and bad legislation" (Radaelli 1999, 3). Such complaints reflect both unease about the tendency of the European Commission to be more of a "policy entrepreneur" than many had expected as well as unease about its technocratic tendencies, including its tendency to produce regulations aiming more at efficiency than at resource redistribution (Radaelli 1999, 6).

The Commission receives very broad guidance from the European Council composed of the Heads of Government of the member states, but the European Council does not itself legislate or propose legislation.

The legislative branch of the EU governance structure is composed of two units: the Council of the European Union (not to be confused with the European Council)—an upper house composed of ministers from each state, which has the unique task of setting guidelines for the Common Foreign and Security policies and ratifying international treaties negotiated by the Commission, and a democratically elected lower house—the European Parliament—which appoints members of the Commission and acts with the Council of the EU to establish a budget and to enact, modify, or reject legislation which has been proposed by the Commission. There is a complex negotiated system which grants the EU exclusive competence to make laws and agreements in some areas. There are some areas in which member states may only legislate if the EU has not done so; some in which the EU may not pre-empt member States; and yet other areas in which the EU can only act to support member State actions.

There is general agreement that in some relatively narrow domains such as the establishment of monetary policy or the establishment of radiation exposure standards for workers in nuclear facilities, technical expertise is so central that it makes sense for experts to play the central role in formulating as well as implementing policies, whereas in broader areas there should be a greater role for elective bodies. The EU has chosen to deal with the issue of balance between technocratic policy formation and democratic governance by creating structures that delegate narrowly defined policy formation—that is, establishing monetary policy—to institutions—that is, the European Bank in concert with representatives of the National Banks—which are almost completely separated from political processes. In the slightly broader areas in which the European Commission has major responsibility for formulating and implementing legislation, its views must be ratified or enacted by the Council of Europe, the members of which are in turn responsible to their national governments.

While some advocates of more direct democratic governance are uncomfortable about these structural arrangements they seem workable to most advocates of continuing national sovereignty, which are in the majority among European voters. It is generally admitted, however, that because legislation must be initiated within the European Commission which has a tradition of technocratic decision making, EU policies are subject to more technocratic influence than those of the individual member states.

Methods of Linking Expertise to Democratic Politics (2): Citizen Participation in Technology Assessment

Formal technology assessment emerged as a tool of policy advice in the United States during the 1960s in large part because of public distrust associated with such projects as the anti-ballistic missile initiative and the support of commercial supersonic aviation. But in spite of the fact that it was initiated in response to broad public unease about issues in which technology played a central role, it developed within organizations such as the President's Science Advisory Committee (PSAC) panels and the subsequent and relatively short-lived congressional Office of Technology Assessment (OTA) as an expert-driven process on the grounds that "non-experts are ill equipped to comment on complex technical matters" (Sclove 2000, 33). Moreover, although OTA reports were made available to the public, their primary audience was members of Congress, or more often, their staff members, who were sometimes scientifically and technologically knowledgeable. PSAC panel reports were covered by executive privilege and generally not made available either to Congress or to the public. As a consequence, the results of technology assessments in the United States rarely came to inform public debate except in cases where PSAC panel members blew the whistle when they felt that the president had intentionally misrepresented their advice (See p. 154).

More recently, however, advocates of greater public participation in policy setting have promoted the view initially developed by Morris Cooke and the early New York Bureau of Municipal Research that one of the most important functions of experts is to inform the public about policy-relevant issues so that the public can both play an informed and significant role in setting policy and hold administrators responsible for efficient implementation of those policies.

In 1987, the Danish Board of Technology was charged both to "initiate comprehensive assessments of the possibilities and consequences of technological developments for society" for the Danish Parliament, and to "support and encourage public debate on technology" (Joss 1998, 7). In response to its charge, it initiated a process involving ordinary citizens in "Consensus Conferences" to assess the possibilities associated with technologies and to report their findings to both Parliament and the general public. In spite of early concerns on the part of technology producers and conservative politicians that such conferences would negatively impact Danish competitiveness in technology-driven fields, the process is now widely supported by virtually all stakeholders, including the Danish Federation of Industry and the two major Danish biotechnology companies (Joss 1998, 19). This is so at least in part because Consensus Conferences have identified public concerns early enough in the Research and Development process that it has allowed firms to anticipate problems and refocus R&D to avoid investing in unmarketable products and processes (Sclove 2000, 39). In addition, minority party members of Parliament see the conferences as ways to "come up with additional alternat[ives]" to Ministry-sponsored policies (Joss 1998, 8).

Topics for consideration are identified by members of Parliament or by the Technology Board as ones which are likely to come up for legislative deliberation, which are likely to be of broad social significance, and which have a significant technical component. A few of the early topics included genetic engineering, food irradiation, air pollution, human infertility, and sustainable agriculture. The Danish Board of Technology then establishes a steering committee chaired by a member of the Board's staff and consisting of relevant experts from academia, industry, and public interest organizations.

While the steering committee chooses appropriate background readings and arranges for additional experts to make presentations to the participants, the board solicits volunteers for participation by advertising in local newspapers throughout the country. Each candidate is asked to submit a one page statement explaining why they want to participate. For each of the nineteen conferences held by the year 1999, the Board received between one hundred and two hundred applicants, from whom it chose about fifteen, representative of the demographics of the population with respect to age, gender, etc. Each group contained homemakers; office and factory workers; public service workers, such as garbage collectors or street maintenance workers; as well as university-educated professionals. The only common characteristics of participants were that none were deemed to have a specific prior financial or personal interest in the topic, and all were sufficiently literate and interested to apply.

At an initial weekend meeting the citizen group, led by a trained facilitator, discusses an expert-prepared background paper commissioned by the steering committee and formulates questions that they wanted addressed at a public forum. The steering committee then assembles an expert panel to respond to those questions—which always includes not only technical issues, but also ethical and social ones—and to represent important stakeholder groups. After a second weekend meeting to discuss more readings, to refine their questions, and to approve or modify the panel organized by the steering committee, the lay group participates in a four-day public event open to the media, members of Parliament and interested citizens. On the first day of the public forum, the expert panel members and stakeholder representatives each give 20–30-minute presentations which include responses to the lay group's initial questions. On the second day, having met to discuss what they heard, the lay group publicly cross-examines the experts to draw out conflicting perspectives and fill in gaps. On the third day the lay group drafts its report with secretarial and copyediting support, but without content-related help from the Board. On the final day, after offering experts a brief opportunity to correct misstatements of their testimony, the lay group presents its report verbally at a national press conference and in written form to the Parliament, which makes it available to the public to stimulate additional public discussion (Sclove 2000, 34–36).

Evaluating the success and potential generalizability of the Danish model is complicated in part because the Danish Consensus Conferences took place in a nation which has historically seen more than average public interest in and awareness of scientific and technological issues as well as more participatory democracy (Joss 2002, 223–24; Andersen and Jaeger 1999, 333–34; Einsiedel et al. 2001, 85). The Danish Board of Technology has had no trouble finding volunteers to take part in Consensus conferences while a project director at U.S. Office of Technology Assessment claimed that "citizens simply refused to participate" in its report review process (Sclove 2000, 42). On the other hand, NGOs in the United States and elsewhere have had little trouble in getting volunteers representing a broad cross section of citizens for Consensus Conferences. OTA's problem may have been that their report review process held little appeal because its format offered no social engagement for volunteers and no assurance that their views would be heard.

Simon Joss carried out an assessment of the first thirteen Danish Consensus Conferences during 1995–1996, giving questionnaires to members of the Danish Parliament and to the general public, and doing in-depth interviews with Parliament members from each of the five major parties (Joss 1998). Parliamentary questionnaires were sent shortly after an election, when 30 percent of the MPs had not yet served, which may have skewed responses to some questions. 75 percent of the MPs knew of the existence of Consensus Conferences. Unsurprisingly those in left of center parties were more knowledgeable than those in parties identified as right of center. Fifty percent had attended one or more Consensus Conferences. Twenty-one percent read all Consensus Conference reports, and an additional 59 percent read the reports on topics about which they had a special interest.

Twenty-nine percent saw Consensus Reports as leading directly to Parliamentary actions, including a prohibition against using information from genomic research in connection with employment or pension schemes. But 47 percent agreed that they were most important indirectly, as the following comment from a member of the Social Democratic Party suggests:

> [Parliament] does take the reports of consensus conferences into account, especially when they touch upon ethical or social issues. For example, the consensus conference on the future of private motoring had an impact on Parliament. . . . But the reports are used more indirectly, they build part of your background knowledge and can be used when you need to find out about a specific topic. . . . Of course, if a conference is organized at a time when Parliament considers the same issue, then you can get direct referring in parliamentary debates, but that is not always the case. (Joss 1998, 15)

Forty-three percent thought that the most important impact of Consensus Conferences was in stimulating public debate on the issues. Perhaps most impressively, though the initial legislation establishing Consensus Conferences had passed by a bare majority in 1985, ten years later 75 percent of the MPs, including a majority of the members from the Conservative Peoples Party, which had opposed the original legislation, supported continuation of the program.

In-depth interviews of members from all parties elicited favorable responses. The following comments from a member of the Socialist People's

Party illuminates two of the reasons that MPs found Consensus Conferences particularly valuable to legislators:

> as a politician you're always a bit in a dilemma: say you need information about a certain issue, so you go to experts—but experts have their own interests to protect and just say 'go on, develop this technology.' If you go and ask people in the street, they respond immediately to your question, but often they do not know what you or they themselves are talking about, and they don't give an informed opinion. If you consult the media then you often get very distorted accounts that are not at all representative of what the public think. Therefore we find it very useful to get a panel of ordinary people to spend some time at a conference listening to the different experts, questioning them, and then—based on their newly acquired knowledge, but at the same time still on their personal feelings and attitudes—stating what they think we should do. . . . Parliament should not act on scare stories, nor only on experts who give their own opinion. We were sent [to Parliament] by the people, so we should have to react on public opinion, but hopefully on informed public opinion (Joss 1998, 11–12)

First, this statement indicates a skepticism regarding the objectivity of expert advice which is more widespread in Europe than in the United States. Second, and more importantly, it suggests that what politicians need or want even more than technical facts is an understanding of the social perceptions of people who are informed regarding issues with a technological orientation. According to Joss, "[All] interviewed Members of Parliament were clear that technology assessment has to address, and be oriented towards, socially relevant issues, and not only be based on purely scientific assessment; hence the need for participatory mechanisms" (Joss 1998, 13).

Given the fact that MPs thought that the most important function of Consensus Conferences was to stimulate and inform public awareness and debate, what evidence is there that it has done so? In a questionnaire distributed to members of the public shortly before and shortly after a Consensus Conference on gene therapy, it was clear that relatively few members of the general public recognized the term "Consensus Conference"—17 percent before and 21 percent after, but 36 percent and 45 percent, respectively, among college educated citizens—but their awareness of the topic, based on increased newspaper, radio, and television coverage after the conference increased by a statistically significant (at the 99 percent confidence level) 5 percentage points, from 61 percent to 66 percent. Anecdotal evidence from MPs suggests that Consensus Conferences have stimulated local governmental bodies and private organizations to initiate debates and discussions (Joss 1998, 16–17). Thus while there is some indication that Consensus Conferences play a role in informing and activating public debate, it is less impressive than the evidence of influence on parliamentary deliberations.

Because of its successes, the Danish Consensus Conference model has been adopted and adapted by several non-Danish governmental organizations responsible for legislation and/or regulation, by public institutions with educational roles to play, and by private groups wishing to promote greater public roles in policy making. Finland, the Netherlands, and Switzerland have formally adopted analogous processes to consider possible regulatory statutes or executive orders. In order to give its decisions more democratic legitimacy the European Commission has begun to use a second and related Danish strategy, the "Scenario Workshop," which involves citizen panels in evaluating and criticizing competing policy scenarios created by experts (Andersen and Jaeger 1999, 338). The Science Museum in London and the Deutsches Hygiene Museum in Dresden have sponsored consensus conferences on plant biotechnology and human genetic testing, respectively, not because they had special responsibility in these areas, but because they wanted to explore the possibilities of bringing together experts, stakeholders and the public to discuss policy options in a public setting. Finally, in Austria, Australia, Canada, France, Japan, New Zealand, South Korea, and the United States, Consensus Conferences have been sponsored by NGOs wishing to promote more participatory democratic practices (Joss 2002, 225; Einsiedel et al. 2001, 83–98).

The Consensus Conference model has four major advantages over the confrontational model of interaction between expertise and democratic processes and one central advantage over the structural system of the EU, especially for those who advocate elements of direct participatory democracy. First, it does not depend upon the grassroots self-organization of citizens around a topic, which is often difficult to achieve. Second, and related to the first, it can be a regular feature of representative democratic institutions and does not depend upon an ideology of participatory democracy; so it is much more widely applicable. Third, it can be used in cases which have not yet generated strongly held and relatively rigid public views—that is, it can be proactive rather than merely reactive to some proposed policy. Fourth, Consensus Conferences are intended to involve a representative cross section of the population rather than a group with some specific prior interest in the topic. In that sense it can be seen as even more democratic than confrontation. In contrast to EU structures, which have historically offered no participatory role to ordinary citizens, Consensus Conferences let non-experts play a significant role in policy considerations.

On the other hand, it may be that the Consensus Conference model is uniquely suited to a relatively well-informed and reflective civic culture such as that in Denmark. Furthermore the Consensus Conference model, at least as practiced in Denmark, responds only to issues identified as interesting to a political elite, and many of the parameters of discussion are set by the experts on the steering committee and the panel members who they choose to inform the laypersons involved. While the citizen panel may probe beyond the expert-defined domain of interest, it is not totally clear how much their views are predetermined by the specialist information that they are exposed to.

Methods of Linking Expertise to Democratic Politics (3): Research Responsive to Community Needs

The National Institute of Environmental Health Sciences (NIEHS) in the United States has, since 1993, been a leader in fostering programs that involve active collaborations between local communities and health care researchers. Under pressure from the National Academy of Science's Institute of Medicine to "engage the public to a greater extent in informing the process by which NIH sets its priorities" (O'Fallon et al. 2003, 1856), NIEHS held sixteen Town Meetings between 1998 and June of 2003 to provide a context in which state, local, and federal health care professionals, the lay public, and advocacy groups could establish common interests and potential collaborations. Typically these Town Meetings were hosted by community organizations and were set up so that a scientific symposium on some topic by scientific researchers was followed by an open forum in which community members were encouraged to ask questions and make suggestions regarding how the scientific research might be applied to their community needs. Positive results included higher awareness of environmental health issues, greater trust between different stakeholders, and occasionally greater coordination between federal and local governmental agencies or collaborations between local health care centers and grass roots organizations on a variety of projects including screening for asthma and providing safer methods for pest control (O'Fallon et al. 2003, 1858). But agendas were first set by the experts before citizens groups could imagine how to use that expertise.

Since the early 1990s, community-based participatory research (CBPR) or, "systematic inquiry, with the participation of those affected by the problem, for the purposes of education and action or affecting social change" (Minkler 2010, S81; Ottinger 2013), has been growing as a strategy for integrating scientific expertise and democratic participation, especially in connection with public health issues both in the United States and beyond. This move has been encouraged by both private and governmental funding agencies, who view partnerships between community-based organizations and health care experts as leading to policies that are more responsive to local issues and more likely to be accepted by the relevant populations. In addition, such partnerships build confidence and capacity to understand and implement new policies in the future (O'Fallon et al. 2003, 1855). From the standpoint of democratic participation in policy making, CBPR has the great advantage over the methods discussed above in that it involves community members in identifying and formulating problems and policy options from the beginning rather than placing them in a position in which they must react to expert-set agendas and suggestions.

A study initiated in 2004 identified some eighty examples of CBPR in the United States associated with the issue of health disparities alone (Minkler 2010, S81). The issue of health disparities highlights a special concern which relates to democratic ideals, for more than many policy issues it obviously emphasizes both distributive justice and procedural justice. The first aims at equalizing exposure to environmental hazards and access to resources including safe recreational areas, healthy food, and health care facilities. The second aims at processes which equalize the role of all segments of the population, including those with low incomes, those in rural regions, and those of different ethnicities and sexual orientations in having an effective voice in decision making that affects their lives. While distributive justice presents no special challenge to expertise, procedural justice almost certainly can, because expertise is unevenly distributed in society and is concentrated among urban, wealthy white populations in the United States. Thus experts are relatively insensitive to the interests and abilities of the poor and minorities. That problem is reduced in CBPR projects in which expertise is intended to respond to the self-defined needs and wishes of local stakeholders.

In a typical CBPR case a local activist group identifies some health problem it hopes to solve. For example, a nonprofit youth empowerment organization, Literacy for Environmental Justice (LEJ) in the largely poor Bayview section of San Francisco identified access to healthy food as a serious problem. The last major chain supermarket had pulled out of the community, leaving small neighborhood markets as the only local sources of food. These markets provided very little fresh produce, and what they did provide was costly and of poor quality. LEJ sought and received funding and support from The San Francisco Department of Health under its Tobacco Free Project. High School students were taught the department's Community Action Model which included developing "research skills for identifying a problem and its root causes and contributing factors, [techniques for] gathering data, developing systems- or policy-oriented solutions, and evaluating [the success of the implemented solution]" (Minkler 2010, S83). In particular they were taught store shelf diagraming, which allowed them to identify the amount of shelf space allocated to different types of products in local stores.

The students found that 26 percent of shelf space was allocated to alcohol and tobacco, 17 percent to sodas and other beverages, 40 percent to packaged foods, the vast majority of which (80–90 percent) were products of Kraft and Nabisco, both subsidiaries of tobacco companies, and only 2 percent to 5 percent for produce and other healthy foods. Furthermore, they determined that in order to reach the nearest full service supermarket, the average resident of Bayview took an hour and required riding on three different busses.

Finally, the students interviewed store owners to determine what kind of incentives could lead them to stock healthier foods, and a U.C. Berkeley graduate student in Public Health was recruited to do an economic feasability study to determine the likely outcomes to store owners if the incentives were offered and they did, in fact, stock healthier food.

In forming its policy recommendations, LEJ and its partners decided to opt for a voluntary program, in part because, "they didn't want to go into the neighborhood and say 'we're another group telling you what you should be doing'" (Minkler 2010, S84). Working with the local Redevelopment Agency and other municipal agencies as well as a very charismatic local supervisor, they got agreement from the Redevelopment Agency to offer facade improvement and free advertising to participating stores, and other agencies arranged for discounted loans and the provision of energy efficient appliances in the future.

A pilot project with a single store was undertaken in 2003. Sales of fresh produce, which had accounted for 5 percent of sales before, jumped to 15 percent of sales within seven months, and alcohol sales dropped from 25 percent to 15 percent. What is in many ways more important, with the new balance of products, profits at the store were up 12 percent after four years during a period of recession that left profits for most local businesses reduced. Four additional stores joined the program by 2007, and five additional stores (for a total of ten out of the eleven local grocery stores in the Bayview neighborhood) were scheduled to join before severe recession eliminated the Redevelopment Agency and the incentives that it had offered participating stores (Minkler 2010, S84). In the mean time, the success of the Bayview pilot project led the California Legislature to pass AB 2384, creating a statewide demonstration program based largely on the Bayview model in 2006.

The Bayview project highlights both the potentially great strengths of CBPR and some of the limitations in linking expertise with democratic politics. First, because CBPR projects tend to have strong buy-in by locally influential groups and because the aims of the research are locally defined, the results usually generate very little resistance within the community. In addition, local community activists are likely to be sensitive to issues that could possibly generate opposition when policies are proposed. Because the research is usually undertaken by community members, subjects tend to be more willing to participate than they would be if outside experts were doing the data gathering. Furthermore, because problems identified in one locality usually exist elsewhere, CBPR projects can often be scaled up for broader application.

On the other hand, successful CBPR projects seem to require the prior existence of a strong and well respected local advocacy group—a requirement which is less often the case in poorer and non-white communities than in middle class ones, constituting a de facto barrier to procedural justice. In addition, they frequently depend on external political contingencies—in the Bayview case, the election of a supportive and charismatic supervisor who could lean on the Redevelopment Agency for help. This last feature is particularly important because CBPR policy recommendations usually require funding from outside agencies over which the initiating groups rarely have significant leverage.

Finally, CBPR is particularly appropriate in connection with health related issues because they tend to involve high levels of community awareness to start with. Similarly, one could imagine other issues, such as traffic control or local educational quality, being of such local importance that they might generate community initiated research proposals. CBPR, however, is an unlikely approach to a huge range of important but non-local issues such as as national security, alternative energy sources, global warming, monetary policies, stem cell research, and so forth.

An only slightly different version of science and technology in the service of a specific community has been occurring without government encouragement in connection with NGOs and other organizations often formed around orphan diseases or disease epidemics. In the mid-1950s, for example, the family of a boy dying from a rare form of muscular dystrophy in France established a network of families in the same position. They exchanged practical information about care giving and began to collect information about the symptoms and temporal progression of the disease as well as to track articles in specialist journals. When their son died in 1958, his parents established *The Association Française contre les Myopathies* (AFM) which collected data, promoted research on neuromuscular diseases, and raised awareness of such diseases. In time they became so adept at raising money for research that by 2005 they had established two different research institutes and generated more than 104 million euros annually to support their activities (Bucchi 2006, 49–50).

Another organization in which citizens initiated their own data collection which led to important consequences was the result of unusually high rates of leukemia among children in Woburn Massachusetts during the 1980s. In spite of health officials' assurances that there was no reason to worry, parents discovered that chemical dumping was occurring nearby. They monitored cases of leukemia over five years and then turned over their results to scientists from MIT, who eventually identified trichloroethylene as the culprit (Bucchi 2006, 51). In these cases citizens initiated the research by doing their own data collection before experts entered the picture, belying the notion, widespread among scientists, that lay persons are too ill educated to appreciate and participate in scientific activities without technocratic direction (Polanyi 1962, 67, 72; Levitt and Gross 1994, B1–B2).

Rethinking the Character and Adequacy of Scientific and Technical Expertise

In almost all of the discussions of the relationships between democratic politics and expertise above I have implicitly accepted the notion that a somehow perfected expertise—an expertise whose rationality was truly unbounded—might accurately and objectively represent *REALITY* in all of its complexity, including its psychological, emotional, and broadly cultural dimensions. Similarly I have implicitly suggested that a perfect democratic process—whether participatory or representative—might effectively generate a consensus regarding the general good. Together, then, a perfect expertise linked to a perfect democratic politics might produce a perfectly Utopian society. But since the early 1960s both of these implicit assumptions have been seriously challenged. The situation has been nicely characterized by Mark Brown in his *Science in Democracy: Expertise, Institutions, and Representation:* "Science can help democracy become more reflective—that is, more informed, competent, deliberative—but science is not a mirror of nature, and government decisions cannot perfectly reflect a preexisting popular will" (Brown 2009, 7–8).

Political theorists have never been able to come to agreement regarding the capacity of democratic politics to satisfy the needs and desires of all citizens in a state; but at least until the mid-twentieth century, many philosophers of science as well as most scientists and technocrats probably felt that science and reason could at least achieve representations of the world that continually approached an accurate reflection of *REALITY*. Since the nearly simultaneous publication of Thomas Kuhn's *The Structure of Scientific Revolutions* (1962) and Karl Popper's *Conjectures and Refutations* (1963), however, most philosophers of science, most of those associated with the new fields of Social Studies of Science and Technology or Science, Technology, and Society (STS), as well as a very small number of scientists and an even smaller number of technocrats have abandoned this claim. They have come to understand scientific progress without the assumption that its goal is the perfect representation of some [ultimately unknowable] reality.

Science and technology do undoubtedly progress; but they do so by moving *away* from areas of ignorance and impotence rather than toward some perfect knowledge or technology. Like other evolutionary processes, science and technology continuously adapt to specific environments and build upon prior stages in their development; so the traditional notions of disinterested and objective knowledge must be abandoned in favor of an awareness that scientific knowledge is inevitably situated knowledge—situated both within a local physical and social environment and within a historical tradition. Such awareness does not necessarily challenge the claim that at any particular place and time scientific knowledge is the most reliable knowledge that we have access to regarding the phenomenal world nor that technical expertise provides a powerful guide to action (Collins & Evans, 2007). It might, however, suggest that claims to the universal applicability of cosmopolitan science and technology should be more modest in the face of local indigenous systems of knowledge and practice that reflect particular environments and experiences—what James Scott has called Mētis (Scott 1998, 311–341), but which we might call indigenous sciences and technologies.

The new way of thinking about the nature of science and technology as *a* form of adaptation to the world might also suggest more modesty regarding the notion that they constitute the *only* effective adaptive strategy. In this regard, David Sloan Wilson, evolutionary biologist and author of *Darwin's Cathedral: Evolution, Religion, and the Nature of Society* has made an argument that seems particularly powerful to me. He writes:

> After Darwin, numerous philosophers and biologists tried to place epistemology on an evolutionary foundation by saying that the ability to know is adaptive (Bradie 1986). Those who did it well survived and reproduced while those who did it poorly were not among our ancestors. This argument has an element of truth; clearly, I need to accurately perceive the location of a rabbit to hit it with my throwing stick. However, there are many, many other situations in which it can be adaptive to distort reality (Wilson 1990, 1995). Even massively fictitious beliefs can be adaptive, as long as they motivate behaviors that are adaptive in the real world. At best, our vaunted ability to know is just one tool in a mental tool kit that is frequently passed over in favor of other tools—just as we observe in all cultures, including our own.
>
> From this perspective, we should expect moral systems to frequently depart from narrow reasoning on the basis of factual evidence. Once this kind of reasoning is removed from its pedestal as the only adaptive way to think, a host of alternatives becomes available. Emotions are evolved mechanisms for motivating adaptive behaviors that are far more ancient than the cognitive processes typically associated with scientific thought. We might therefore expect moral systems to be designed to trigger powerful emotional impulses, linking joy with right, fear with wrong, anger with transgressions. We might expect stories, music, and rituals to be at least as important as logical arguments in orchestrating the behavior of groups. Supernatural agents and events that never happened can provide blueprints for action that far surpass factual accounts of the natural world in clarity and motivating power. These otherworldly elements of religion cannot completely eclipse scientific modes of thought, which are superior in some contexts, but the reverse statement is equally true. (2002, 41–42)

From this perspective, what scientists and technocrats tend to label as "irrational" approaches to policy may none the less sometimes be what is most appropriate. It is, of course, possible that strategies which were adapted to circumstances in the earlier evolution of humans are no longer useful. On the other hand it is also possible that in some circumstances science and reason may be maladaptive, as they seem to have been in the development and implementation of Eugenics policies in the early years of the twentieth century and in the continuing nearly exclusive use of fossil fuels to produce energy for human use at the beginning of the twenty-first century.

Chapter Eight

Conclusion

On the Mixed Blessings of Scientism and Technocracy

I grew up in the Midwest, where my father had a vegetable garden in the backyard. In that garden he grew tomatoes which ripened in the late summer. They had a marvelous red color, irregular shapes and delicious taste. If you tried to cut one with a slightly dull knife its skin would tear and it would be squashed. Now I own a summer cabin in the Sierra Nevada Mountains in Northern California, and when I drive up the central valley in midsummer I often follow trucks hauling commercially grown processing tomatoes to packing plants along the way. When the trucks round a tight curve, a few tomatoes usually fall out of the top of the bins they are being hauled in, and the vaguely pink, perfectly round, objects bounce along the shoulders of the road like small tennis balls. I'm not even sure that if I stopped to pick one up it would be bruised. Moreover, when I buy commercially grown eating tomatoes in the store, most of them look much like the processing tomatoes, and all have a firm texture and relatively bland taste. Now I never simply eat tomatoes with a little salt on them at a meal, and the frequency with which I eat anything, including BLTs, containing eating tomatoes is very low because they just do not have the same taste and aesthetic appeal as those tomatoes of the distant past.

In another generation or two, except for the handful of wealthy people who can afford the heritage tomatoes grown by a few boutique farmers, no one will even know what they are missing; so consumers will generally be satisfied with what is available. On the other hand, the costs to consumers of the widely used products of tomato processing—ketchup, tomato juice, tomato sauce, tomato paste, canned tomatoes, marinara sauces, bottled salsa, etc.—are all a tiny fraction of what they would otherwise be and they are available all year round. We have traded away a special seasonal gourmet experience, available to some people in the past for a much cheaper and more uniform set of products which are available to much broader segments of the population all of the time.

A similar story can be told about the virtually complete replacement of the sweet, soft, and delicious freestone peach varieties by the much tougher and less tasty cling varieties among commercial peach growers because the former could not survive inexpensive mass transportation to packing plants. Freestones bruise easily; so they demand careful boxing and handling which makes their distribution much more labor intensive and drives their prices up beyond what consumers will pay in the face of much cheaper, though in many ways less satisfying, alternatives.

Now consider briefly the farmer's perspective on the transformation of tomato growing rather than tomato eating. The invention of mechanical tomato harvesters and the scientific development of new, more easily picked and transported varieties of tomato have lowered the labor cost, and indeed, the overall unit cost of producing and transporting tomatoes; but the savings could only be realized for large production units because the capital investment in the machines could only be supported by large farms. As a consequence, the mechanical tomato harvester was one among many agricultural innovations that transformed farming in the United States from an industry dominated by small, family-run enterprises in the mid-nineteenth century to one dominated by huge agribusiness-owned properties by the mid-twentieth. This process reduced the demand for full-time agricultural workers, driving up rural unemployment; and it virtually eliminated a whole way of life that had provided a special satisfaction to millions of Americans. In addition, because mechanical harvesters work best on flat land that is dry during the harvesting season, they were instrumental in driving up the price of land in California's central valley, which was both flat and dependant on irrigation for water during the summer growing season. Moreover, to further increase the production of tomatoes and justify using the increasingly expensive land in the central valley for tomato production, more fertilizers and more irrigation were demanded, creating potential health dangers, diverting water from alternative uses and making water an increasingly expensive commodity for all residents of Northern California. In addition, the monocropping of the highest productivity variety of tomato led to the need for ever greater amounts of pesticides and herbicides to head off the incursion of ever more resistant bugs and weeds.

How did this happen? Why should we care? And what does this have to do with the growth of scientism and technocracy?

The transformation of tomato growing in the central valley of California directly involved technocratic attitudes on the part of the U.S. Department of Agriculture (USDA), which decided to fund work on the mechanization of agriculture to increase agricultural productivity after WWII. As part of this government sponsored push for increasing mechanization, the UC-Blackwelder tomato harvester was developed at UC Davis in 1949 by the agricultural engineer Coby Lorenzen working with vegetable crop researcher Jack Hanna (Coatney 2006). Though the machine had tremendous potential for making tomato harvesting more efficient, it tended to bruise the tomato varieties grown at the time. Hanna set about creating a tomato variety that was both more uniform in size and shape and vastly tougher than traditional varieties so that the tomatoes were more suitable for mechanical harvesting even though they were less tasty. In the mid-1960s tomato harvesters and the new tomato variety were introduced into commercial farms in California's central valley, and within five years they had totally transformed the tomato-growing business. Now, at the beginning of the twenty-first century, California grows 95 percent of all processing tomatoes in the United States (Coatney 2006).

Political opposition based on the loss of jobs led the USDA to withdraw its support for mechanization in the sixties, but it was too late to influence the use of the UC-Blackwelder harvester. Moreover, by 1962, the economic incentives to increase productivity coming from private corporations continued to promote mechanization efforts in agriculture.

The story of mechanical tomato harvesting illustrates a large number of issues connected with scientistic and technocratic attitudes, especially as they have been manifested in the tradition derived from and associated with Scientific Management. The primary goal of technocratic decision making has been to maximize the public good by increasing efficiencies of production—initially the production of material commodities, but subsequently, the activities of private businesses and of public agencies whose products are often intangible, for example, clean streets, education, or reduced infant mortality. Mechanical tomato harvesting represents or symbolizes the dramatic successes of technocratic decisions in terms of this primary goal because it dramatically drove down the costs of tomato growing and served the public by making tomato-based products available year round as inexpensive mass marketed commodities. Moreover, it illuminates how thoroughly the drive for efficiency and productivity has come to inform a huge number of decisions we make about activities in our lives.

On the other hand, the story illustrates the often unconsidered or even unconscious way in which the emphasis on productivity discounts unquantifiable and subjective aspects of experience—things like the tastes, textures, and colors of the objects that we live with—as it focuses on the measurable and objective—that is, the tonnage of tomatoes produced per dollar of cost. When, for example, Morris Cooke introduced the student contact hour as a measure of educational efficiency he ignored the quality of those hours of contact. Similarly, today when we measure the effectiveness of education by the scores which students achieve on standardized tests of content, it is not at all clear that we are measuring what is most important about education, which may be creating an enthusiasm for learning and a capacity for discovering knowledge independently on the part of students. From the perspective of traditional technocrats, the product of education is knowledge of content, which can be relatively easily measured. In their view, the wish to have students love learning is irrelevant and irrational except to the extent that it might be used to increase the content of students' knowledge; but I suspect that many of what I would judge to be the best educators would argue that love of learning is a legitimate goal for its own sake rather than as an instrument for accomplishing something else.

I have used an extension of Herbert Simon's concept of bounded rationalities to suggest that historically, the bounds of technocrat's rationality have been limited relative to those of other groups, whether they be members of the military and political leadership during the Cold War or Nimby's today. While a technocratic perspective had been critically important in WWII and in developing military hardware and tactics during the Cold War, the tendency for military personnel to adopt the rationality of technocrats may be restricting their abilities to think and act strategically in the contexts which they face today (pp. 92–94). The Nimby cases discussed above illustrate an opposite problem. While technocrats hope to be objective and unbiased in developing policy advice, and while ideally they aim to serve the general good, they often have difficulty in exploring an appropriate range of policy options because they must limit considerations to those which fit within the limited rationality of some clients. In such cases, opposition to technocratic views may force consideration of options closed by clients, though it is likely to be seen as irrational and obstructionist by the technocrats (pp. 135–136).

The tomato story also suggests the way in which efficiencies in the production of a given commodity may have the effect of reducing human choices by driving alternative products out of the market. One might argue that because there are now ten different brands of ketchup available in the supermarket at prices that make them all accessible to virtually everyone, people have more choices by virtue of increased productivity. But Herbert Marcuse, among others, argues that such is merely an illusion of choice rather than a real choice. Consumerism, which is the flip side of the emphasis on productivity, Marcuse argues, controls us and actually deprives us of real choices (Marcuse 1964)—in the tomato case, because all brands are made from the same varieties of relatively bland tomatoes.

Moreover, because greater efficiency is often associated with mass production and hence mass markets for a product, consumers with special desires or needs often find themselves with no suitable choices. I cannot find Freestone peaches in my local supermarket at any price, for example. And the mass machine manufacturing of shoes dramatically reduced the cost of shoes to the consumer in terms of labor hours necessary to purchase a pair of shoes; but no machine-manufactured shoes are made with heels that are about two standard deviations narrower than the norm, and mechanical shoe manufacture has effectively driven custom shoe manufacture beyond the reach of all but the wealthiest members of society, leaving people with feet that deviate significantly from the norm wearing shoes that do not fit.

The mechanical harvesting of tomatoes story also illustrates the way in which emphases on production have often ignored the impacts on the producers and their ways of life. Whether for good or for ill, American society today, with fewer than 4 percent of the population involved in primary food production, is radically different from what it was in the mid-nineteenth century when nearly a third of the population was involved. Furthermore, focus on the efficient production of one commodity has often ignored how that production might affect the costs of some other entities, such as land and water, or of other externalities, including additional health care costs deriving from increased herbicide and pesticide use.

There is some evidence that within engineering, which nurtured the development of technocratic attitudes during the early twentieth century, technocratic thinking has at least become much more sophisticated than it once was. While efficiency of production still remains the major goal in almost all cases, engineers are being educated today to take into account broader social and cultural concerns and to think about the long term as well as short term implications of their projects. At the same time, economists are becoming increasingly dominant within both private and public institutions that depend upon technocratic perspectives, as is illustrated both in connection with the roles played by economists in developing countries such as Mexico and in the staffing at institutions such as RAND corporation (pp. 128–129, 150–151). And it looks as if the education of economists has become narrower as that of engineers has become broader. The more students of economics focus on the scientific side of their discipline—the construction and evaluation of ever more complex mathematical models, for example—the more they ignore economic history, the history of their own discipline, and the connections which economics once had to the other social sciences before Alfred Marshall set up a separate economics program at Cambridge (p. 144).

One consequence of the replacement of engineering-oriented technocrats with economists is that while engineering oriented technocrats were often inclined to think of efficiency in ways that emphasized labor expended per unit of production rather than overall costs of production and to be relatively uninterested in such issues as net profits or trade deficits, etc., modern technocracy is much more focused on economic issues and may often favor labor intensive production methods if cheap labor is available. As a consequence of the growing importance of economics, contemporary technocracy may be even more narrowly focused than early-twentieth-century technocracy, though the focus is slightly different and overall cost effectiveness has replaced labor productivity as the dominant measure of efficiency. There is even some danger that just as the American military was forced to become engineering oriented during the Cold War, many fields, including education and health care, are becoming more economics oriented today.

There are a few signs that suggest that economically oriented technocratic decisions are becoming more broad based than they were as late as the 1970s and that they may become even more so in the future. The trend toward transdisciplinary studies of policy concerns offers the possibility that far fewer important contextual issues will simply be ignored and allowed to slip through the cracks. Equally importantly, there are important subfields within economic research today—including experimental economics and evolutionary economics—that are not yet widely integrated into curricula but which suggest that economists are becoming more self reflective about the notions of rationality that they employ. As these subfields expand into curricular emphases we might hope to see the kind of broadening of economics education that has been happening during the last two decades in engineering.

The tomato growing and eating example finally exemplifies a common relationship between technical expertise and both wealth and democracy during most of the twentieth century. The USDA promotion of mechanized farming was neither a response to consumer demand for more and cheaper, but less tasty, tomatoes nor a response to the demands of traditional small farmers to make farming more efficient but more capital intensive. It was a response to a near obsession with efficiency and productivity on the part of agricultural experts and large agribusinesses. While there were unquestionably political elements involved first in the promotion and later in the withdrawal of USDA support for mechanization, the new technologies were invented by agricultural scientists and engineers and implemented by large growers with little if any consultation with either the bulk of consumers or the bulk of farmers that they would impact and with little if any consideration of how it would impact the lives of either group.

In general, technocratic decisions from the time of the earliest Scientific Managers have favored science-based technological innovations, especially the invention of new machines and processes, as a way to increase efficiency and effectiveness. As long as technocratic decisions failed to consider the distribution of goods and services made widely available as a consequence of technological innovation, the tendency for that innovation has been to increase the concentration of wealth because it has favored those with capital assets over those with only labor power to offer. Moreover this has been true, not only within advanced industrial societies, but between those societies and developing nations. In spite of the fact that technocratic decisions have largely been well intentioned and aimed at improving the general welfare of citizens of the developing world, perhaps their greatest impact has been to exacerbate existing inequalities both within developing nations and—except for the Asian Dragons and Brazil—between the advanced industrial world and developing nations.

Though Scientific Management eventually came to embrace workplace democracy, it did so largely because it recognized that shop level workers had special expertise to offer. For the most part, neither scientists nor technocrats have acknowledged such special expertise on the part of the general public; rather they have usually seen the public as at best uninformed and at worst irrational (pp. 135–136). As a consequence, it has rarely been the case that technocrats have felt the need to consult or consider public sentiment because they feel more competent to understand the general good than members of the general public. In the tomato growing case, for example, at no time did the USDA or the agricultural engineers and scientists at UC Davis even imagine that it would be a good thing to consult a group of citizens regarding the introduction of the new tomato harvesting technology. As Scientific Management was appropriated within totalitarian regimes in Germany, the Soviet Union, and in China, this tendency of technocrats to ignore or consciously discount the wishes of the population in favor of their "rationally" determined policies has been particularly obvious.

Beginning in the late 1950s members of the U.S. government's executive branch began to think that the assessment of the impacts of new technologies might be a governmental responsibility, and in the early 1960s the U.S. Congress established a much admired but short lived Office of Technology Assessment. But even then, technology assessment remained the domain of experts. In the 1960s, citizen groups did begin to band together to resist the implementation of some technologies when they feared a local negative impact. Nuclear power plants and toxic waste disposal, for example, generated strong public responses. While such direct democratic opposition has sometimes led to "better" decision making viewed in hindsight and from the perspective of the general good, it has rarely successfully expanded the criteria for decision making beyond that set by experts, including those mobilized by opposition groups. Only beginning in the late 1980s and starting in Denmark have techniques such as consensus conferences (pp. 163–168) begun to seriously involve citizen's groups in bringing public concerns regarding the impacts of new technologies into the center of legislative decision making before the implementation has become a *fait accompli*. Even more recently, a few experiments, primarily connected with public health, have begun to allow the participating public to initiate consideration of issues with a technical component and to call upon experts to serve rather than direct their interests (pp. 168–171).

The expansion of technology assessment from a citizen perspective as well as the emergence of community directed initiatives are very encouraging as new ways to integrate technocratic and democratic tendencies, but the implementation of programs which incorporate them is very slow. At the same time, in virtually all nations and in supra-national organizations, critical decisions, especially decisions regarding economic policy, are now made by technocrats whose responsibility to any democratically elected body is very indirect. Moreover, the growing independence of some technocratic influences from democratic oversight comes at a time when many philosophers and students of science and technology are challenging the very possibility that technical knowledge and technocratic policy advice can be as disinterested and objective as their advocates have long claimed. It is, for example, increasingly clear that the claims of contemporary neoliberal, "scientific," economics incorporates a wide range of assumptions that undermine its claims to objectivity.

Engineering education has recently shown significant efforts to broaden its foundations and to move beyond traditional technocratic emphases to incorporate more extensive contextual considerations, but the same is not true of economics education, which seems to be becoming ever more technocratic as economists increase their influence in both private and public policy decisions. Finally, recent public calls to strengthen STEM education to increase economic competitiveness threaten to reverse any movement toward broadening the bases of technical education of any kind. They remind one entirely too much of George Babbit's lament that "there's a whole lot of valuable time lost even at the U., studying poetry and French, and subjects that never brought in anybody a cent" (Lewis 1922, 65).

Bibliography

ABET, 1998. *Engineering Criteria 2000, 3rd Edition.* Baltimore: Accreditation Board for Engineering and Technology, Inc.

———, 2011. *Criteria for Accrediting Engineering Programs, 2012–2013.* Online version at http://www.abet.org/wp-content/uploads/2015/04/accreditation-policy-and-procedure-manual-2012-20131.pdf Acessed on October 17, 2015.

Adas, Michael, 2006. *Dominance by Design: Technological Imperatives and America's Civilizing Mission.* Cambridge, MA: Harvard University Press.

Afanasyev, V. G., 1971. *The Scientific Management of Society.* Moscow, USSR: Progress Publishers.

———, 1975. *The Scientific and Technological Revolution—Its Impact on Management and Education.* Moscow, USSR: Progress Press.

Alcala, Angel C., 1999. "Higher Education in the Philippines," *Philippine Studies, 47*: 114–28.

Alexander, Jennifer Karns, 2008. *The Mantra of Efficiency: From Waterwheel to Social Control.* Baltimore: Johns Hopkins University Press.

Allum, Nick; Patrick Sturgis; Dimitra Tabourazi; and Ian Brunton-Smith, 2008. "Science, Knowledge, and Attitudes Across Cultures: A Meta-Analysis," *Public Understanding of Science, 17*: 35–54.

Andersen, Ida-Elisabeth, and Birgit Jaeger, 1999. "Scenario Workshops and Consensus Conferences: Towards More Democratic Decision-Making," *Science and Public Policy, 26*: 331–40.

Anderson, Gordon, 1985. "Ideological Convergence of the US and USSR," *International Social Science Review, 60*: 19–34.

Andrea, A.J. and J.H. Overfield, 1990. *The Human Record: Sources of Global History. 2 vols.* Boston: Houghton Mifflin Company.

Aronova, Elena, 2011. "The politics and contexts of Soviet science studies (*Naukovedenie*): Soviet philosophy of science at the crossroads," *Studies in East European Thought, 63*: 175–202.

———, 2012. "The Congress for Cultural Freedom, *Minerva*, and the Quest for Instituting 'Science Studies' in the Age of the Cold War," *Minerva, 50*: 307–37.

Aslanbeigui, Nahid, and Michele Naples, 1997. "The Changing Status of the History of Thought in Economics Curricula," in Aslanbeigui, Nahid and Young Back Choi, eds., *Borderlines of Economics: Essays in Honor of Daniel R. Fusfield.* London: Routledge.

Bailes, Kendall, 1974. "The Politics of Technology: Stalin and Technocratic Thinking among Soviet Engineers," *The American Historical Review, 79*: 445–69.

———, 1977. "Alex Gastev and the Soviet Controversy over Taylorism, 1918–24," *Soviet Studies, 29.* 373–94.

———, 1978. *Technology and Society under Lenin and Stalin*. Princeton, NJ: Princeton University Press.
Banta, Martha, 1993. *Taylored Lives: Narrative Productions in the Age of Taylor, Veblen, and Ford.* Chicago: University of Chicago Press.
Bass, Thomas A., 1997. *Camping with the Prince and Other Tales of Science in Africa.* Wakefield, RI and London: Moyer Bell.
Bauer, Martin W., Rajesh Shukla, and Nick Allum, eds., 2012. *The Culture of Science: How the Public Relates to Science Across the Globe.* New York and London: Routledge.
Beine, Michael, Frédéric Docquier, and Hillel Rappaport, 2001. "Brain drain and economic growth: theory and evidence." *Journal of Development Economics.* 275–89.
Beissinger, Mark R., 1988. *Scientific Management, Socialist Discipline, and Soviet Power.* Cambridge, MA: Harvard University Press.
Bell, Daniel, 1999 (1973 1st edition). *The Coming of Post Industrial Society*. New York: Basic Books.
Bendix, Reinhard, 2001. *Work and Authority in Industry: Managerial Ideologies in the Course of Industrialization.* New Brunswick, NJ: Transaction.
Beveridge, Albert J., 1898. *Congressional Record, 56th Congress, Session 1*: 704–12.
Bimber, Bruce, 1996. *The Rise and Fall of the Office of Technology Assessment*. Albany: State University of New York Press.
Brandeis, Louis, 1965. *The Curse of Bigness: Miscellaneous Papers of Louis D. Brandeis,* O.K. Frankael and C.M. Louis, eds. Port Washington, NY: Kennikat Press.
Brown, Mark B., 2009. *Science in Democracy: Expertise, Institutions, and Representation.* Cambridge, MA: MIT Press
Bucchi, Massimiano, 2010. *Beyond Technocracy: Science, Politics, and Citizens.* Dordrecht: Springer.
Burch, David, 1988. "Science and Development: Trends and Outcomes of the Transfer of Technology in the 1980s" in Sardar, 1988.
Bureau of Labor Statistics, "National Employment Hours and Earnings,"—Average Hourly Earnings, 1982 dollars—Data extracted October 13, 1998.
Burnham, John C., 1968. "On the Origins of Behaviorism," *Journal of the History of the Behavioral Sciences, 4:* 143–51.
Butler, Samuel, 1959 (1872). *Erewhon*. London: Dent.
Callahan, Raymond E., 1962. *Education and the Cult of Efficiency: A Study of the Social Forces that Have Shaped the Administration of the Public Schools.* Chicago: University of Chicago Press.
Caoili, Olivia C., 1983. "A History of Higher Education in Science and Technology in the Philippines," *Philippine Social Sciences and Humanities Review, 47*: 302–33.
———, 1986. "A history of science and technology in the Philippines." *Analysis of Conditions for National Scientific and Technological Self-Reliance: The Philippine Situation*. Quezon City: University of the Philippines.
Cech, Erin, 2013. "The (Mis)Framing of Social Justice: Why Ideologies of Depoliticization and Meritocracy Hinder Engineers' Ability to Think about Social Injustices," in Lucena, 2013, 67–84.
Centeno, Miguel Ángel, 1994. *Democracy within Reason: Technocratic Revolution in Mexico*. University Park: Pennsylvania State University Press.
Clarence, Emma, 2002. "Technocracy Reinvented: The New Evidence Based Policy Movement," *Public Policy and Administration, 17, #3:* 1–11.
Clements, Kendrick A., 2000. *Hoover, Conservation, and Consumerism: Engineering the Good Life.* Lawrence: University Press of Kansas.
Coatney, Kathy, 2006. "The Machine that Revolutionized a Harvest," *Ag Alert: The Weekly Newspaper of California Agriculture.* March 15. Online version at http://www.agalert.com/story/?id=554. Accessed on Wednesday, October 30, 2013.
Cobley, Evelyn, 2009. *Modernism and the Culture of Efficiency.* Toronto: University of Toronto Press.
Colander, David and Harry Lambeth, 2007. "Political Influence on the Textbook Keynesian Revolution: God, Man, and Laurie Tarshis at Yale," Online version at http://cat.middlebury.

edu/~colander/articles/Political%20Influence%20on%20the%20Textbook%20Keynesian%20Revolution.pdf. Accessed on Tuesday, November 5, 2013.

Collins, Harry and Robert Evans, 2007. *Rethinking Expertise*. Chicago: University of Chicago Press.

Cooke, Morris L., 1913. "The Spirit and Social Significance of Scientific Management," *Journal of Political Economy, 21*: 481–93.

———, 1917. *How About It?* Lancaster, PA: New Era Publishing Co.

———, 1918. *Our Cities Awake*. New York: Doubleday, Page, & Co.

Currell, Susan and Christina Cogdell, eds., 2006. *Popular Eugenics: National Efficiency and American Mass Culture in the 1930s*. Athens: Ohio University Press.

Cutcliffe, Stephen H., 1990. "The STS Curriculum: What Have We Learned in Twenty Years," *Science, Technology, and Human Values, 15:* 360–72.

Daito, Eisuke, 1989. "Railways and Scientific Management in Japan 1907– 30," *Business History, 31*: 1–28.

Dankwa-Mullan, Irene; Kyu B. Rhee; David Stoff; Jennifer Reineke Pohlhaus; Francisco Sy; Nathaniel Stinson Jr.; John Ruffin, 2010. "Moving toward Paradigm Shifting Research in Health Disparities Through Translational, Transformational, and Transdisciplinary Approaches," *American Journal of Public Health, 100*, Supplement 1: S19–S24.

Del Mar, Donald and Roger D. Collons, eds., 1976. *Classics in Scientific Management: A Book of Readings*. Tuscaloosa: University of Alabama Press.

De Monthoux, Pierre Guillet, 1991. "Modernism and the Dominating Firm—On the Managerial Mentality of the Swedish Model," *Scandinavian Journal of Management. Vol. 7*: 27–40.

Dos Passos, John, 1936. *The Big Money*. New York: Harcourt, Brace, and Co.

Drucker, Peter, 1954. *The Practice of Management.* New York: Harper.

———, 1973. *Management: Tasks, Responsibilities, Practices*. New York: Harper & Row.

Einsiedel, Edna, Erling Jelsre, and Thomas Breck, 2001. "Publics at the Technology Table: the Consensus Conference in Denmark, Canada, and Australia," *Public Understanding of Science, 10*: 83–98.

Engwall, Lars and Celia Pahlberg, 2001. "The Diffusion of European Management Ideas," *CEMP Report No. 17*. (The Creation of European Management Practice)

Enthoven, Alain, 1995. "Tribute to Charles J. Hitch," *OR/MS Today.*

Erickson, Paul, Judy L. Klein, Lorraine Daston, Rebecca Lemov, Thomas Sturm, and Michael D. Gordon, 2013. *How Reason Almost Lost Its Mind: The Strange Career of Cold War Rationality.* Chicago: University of Chicago Press.

Fairfield, John, 1994. "The Scientific Management of Urban Space: Professional City Planning and the Legacy of Progressive Reform," *Journal of Urban History, 20*: 179–204.

Farish, Matthew and P. Whitney Lackenbauer, 2009. "High modernism in the Arctic: planning Frobisher Bay and Inuvik," *Journal of Historical Geography, 35:* 517–44.

Felder, Richard M., and Rebecca Brent, 2003. "Designing and Teaching Courses to Satisfy the ABET Engineering Criteria," *Journal of Engineering Education, 92(1)*: 7–25.

Fischer, Frank, 1990. *Technocracy and the Politics of Expertise: Critical Perspectives on the Management and Policy Sciences*. Newbury Park, CA: Sage.

Fogarty, Ellie, and Michele Naples, 1998. "The Presence of History: A Survey of Articles on the History of Economic Thought and Methodology in the *Economic Literature Index*, 1969–1995," *Journal of Economic Issues, 32*: 224–34.

Fortun, M. and S.S. Schweber, 1993. "Scientists and the Legacy of World War II: The Case of Operations Research (OR)," *Social Studies of Science, 23:* 595–642.

Frazier, Mark, 2002. *The Making of the Chinese Industrial Workplace*. Cambridge: Cambridge University Press.

Freemantle, Michael, 2014. *The Chemist's War, 1914–1918.* London: The Royal Society of Chemistry.

Galbraith, John Kenneth, 1967. *The New Industrial State.* Princeton, NJ: Princeton University Press.

Ganz, Cheryl R., 2000. "Science Advancing Mankind," *Tchnology and Culture, 41(4)*: 783–87.

Gartman, David, 2000. "Why Modern Architecture Emerged in Europe, Not in America: The New Class and the Aesthetics of Technology," *Theory, Culture, and Society, 17*: 75–96.

Gauchat, Gordon, 2012. "Politicization of Science in the Public Sphere: A Study of Public Trust in the United States, 1974 to 2010." *American Sociological Review, 77*: 167–187.

Gauld, Richard, 2000. "Maintaining Centralized Control in Community-based Forestry: Policy Considerations in the Philippines," *Development and Change, 31*: 229–54.

Ghamari-Tabrizi, Sharon, 2005. *The Worlds of Herman Kahn: The Intuitive Science of Thermonuclear War*. Cambridge, MA: Harvard University Press.

Giddings, Franklin, 1901. *Democracy and Empire: With Studies of Their Psychological, Economic, and Moral Foundation.* New York: The Macmillan Company.

Gilbreth, Frank, 1911. *Motion Study.* New York: D. Van Nostrand.

Gilbreth, Frank and Lillian Gilbreth, 1920. *Motion Study for the Handicapped.* London: George Routledge.

Glaeser, Bernard, ed., 1987. *The Green Revolution Revisited: Critique and Alternatives.* London: Allen and Unwin.

Go, Julian, 2007. "The Provinciality of American Empire: 'Liberal Exceptionalism' and U.S. Colonial Rule, 1898–1912," *Comparative Studies in Society and History, 49*: 74–108.

Goodnow, Frank, 1900. *Politics and Administration: A Study in Government.* New York: Russell and Russell.

Gottwald, Jon-Carsten, 2005. "The Regulatory State and Financial Services: An Appraisal of Recent Developments in the European Union," Paper Presented at the EPCR Conference, Budapest, 2005. 1–23.

Green, Philip, 1968. "Science, Government, and the Case of RAND: A Singular Pluralism," *World Politics, 20*: 301–26.

Greenberg, Daniel S., 2001. *Science, Money, and Politics: Political Triumph and Ethical Erosion.* Chicago: University of Chicago Press.

———, 2007. *Science for Sale: The Perils, Rewards, and Delusions of Campus Capitalism.* Chicago: University of Chicago Press.

Guillén, Mauro, 1990. "States, Professions, and Organizational Paradigms: German Scientific Management, Human Relations, and Structural Analysis in Comparative Perspective," Paper presented at the *Center for Advanced Study in the Social Sciences* of the the Juan March Institute, October, 1990.

———, 2006. *The Taylorized Beauty of the Mechanical: Scientific Management and the Rise of Modernist Architecture.* Princeton, NJ and Oxford: Princeton University Press.

Guimond, Serge, 1999. "Attitude Change During College: Normative or Informal Social Influence," *Social Psychology of Education, 2:* 237–61.

Gulick, Luther, and L. Urwick, eds., 1937. *Papers on the Science of Administration.* New York: Columbia University Institute of Public Administration.

Gunnell, John C., 1982. "The Technocratic Image and the Theory of Technocracy," *Technology and Culture, 23:* 392–416.

Guston, David, 1999. "Evaluating the First U.S. Consensus Conference: The Impact of the Citizen's Panel on Telecommunications and the Future of Democracy." *Science, Technology and Human Values, 24*: 451–81.

Haber, Samuel, 1964. *Efficiency and Uplift: Scientific Management in the Progressive Era, 1890–1920.* Chicago: University of Chicago Press.

Hagerott, Mark, 2008. *Commanding Men and Machines: Admiralship, Technology, and Ideology in the 20th Century U.S. Navy*. Phd Dissertation, University of Maryland, College Park.

Halberstam, David, 1972. *The Best and the Brightest.* New York: Random House.

Hallin, Daniel, 1992. "The Passing of the 'High Modernism' of American Journalism," *Journal of Communication, 42*: 14–25.

Happe, Kelly E., 2013. *The Material Gene: Gender, Race, and Heredity after the Human Genome Project.* New York: New York University Press.

Haque, Nadeem U., and Se-Jik Kim, 1995. "'Human Capital Flight': Impact of Migration on Income and Growth." *IMF Staff Papers*, 577–607.

Harding, Sandra, 1986. *The Science Question in Feminism*. Ithaca: Cornell University Press.

Hayden, Thomas, 2005. "Science: Scientists and Bush Administration at Odds," *U.S. News and World Report,* Posted June 30, 2005. Accessed Monday, October 7, 2013.

Hazama, Hiroshi and Jacqueline Kaminski, 1979. "Japanese Labor-Management Relations and Uno Riemon," *Journal of Japanese Studies, 5*: 71–106.
He, Li, 2006. "Returned Students and Political Change in China," *Asian Perspective, 30*: 5–29.
Headrick, Daniel, 1988. *The Tentacles of Progress: Technology Transfer in the Age of Imperialism, 1850–1940.* New York and Oxford: Oxford University Press.
Herf, Jeffrey, 1984. " The Engineer as Ideologue: Reactionary Modernists in Weimar and Nazi Germany," *Journal of Contemporary History, 19:* 631–48.
Herkert, J., 2006. "ABET's Engineering Criteria 2000 and Engineering Ethics: Where Do We Go From Here?" Online Ethics Center for Engineering 6/26/2006, National Academy of Engineering, Accessed: Thursday, September 12, 2013, www.onlineethics.org/Education/instructessays/herkert2.aspx.
Hirdman, Yvonne, 1994. "Social Engineering and the Woman Question: Sweden in the Thirties," *Studies in Political Economy, 44*: 73–94.
Hoffman, Erik P. and Robbin F. Laird, 1985. *Technocratic Socialism: The Soviet Union in the Advanced Industrial Era.* Durham, NC: Duke University Press.
Hofstede, Geert, 1993. "Cultural Constraints on Management Theories," *The Executive, 7:* 81–94.
Holmes, John H.; Amy Lehman; Erin Hade; Amy K. Ferketich; Sarah Gelhart; Garth H. Rauscher; Judith Abrams; and Chloe E. Bird, 2008. "Challenges for Multilevel Health Disparities Research in a Transdisciplinary Environment," *American Journal of Preventative Medicine, 35*: Supplement 1, S182–S192.
Holston, James, 2009. "The Spirit of Brasília: Modernity as Experiment and Risk," in *City/Art: The Urban Scene in Latin America.* Edited by Rebecca E Biron. Durham, NC: Duke University Press.
Hoover, Herbert, 1951. *The Memoirs of Herbert Hoover. Volume 1: Years of Adventure, 1874–1920.* New York: Macmillan.
Hopkins, George B., 1912. "The New York Bureau of Municipal Research," *Annals of the American Academy of Political and Social Science, 41*: 235–44.
Hunter, S., and Leyden K. M., 1995. "Beyond NIMBY: Examining Opposition to Hazardous Waste Facilities," *Policy Studies Journal, 23*(4), 601.
Huxley, Aldous, 1946. *Brave New World.* New York: Harper and Brothers.
———, 1962. *Island: A Novel.* New York: Harper and Brothers.
International Rivers Network, 1993. *Large Dams, False Promises.* (Video) Berkeley California.
Ipsos MORI Social Research Institute, 2011. *Public Attitudes to Science, 2011.* London
Jaschik, Scott, 2012. "Study Tracks Erosion of conservative confidence in science," *Inside Higher Ed* (http://www.insidehighered.com) March 29.
Jordan, John M., 1994. *Machine Age Ideology: Social Engineering and American Liberalism, 1911–1939.* Chapel Hill: University of North Carolina Press.
Josephson, Paul, 1986. "The Historical Roots of the Chernobyl Crisis," *Soviet Union, 13*: 275–99.
———, 1995. "'Projects of the Century' in Soviet History: Large-Scale Technologies from Lenin to Gorbachev," *Technology and Culture, 36:* 519–59.
Joshu, D. And T.D. McBride, 2013. *Transdisciplinary Public Health: Research, Methods, and Practice*. San Francisco: Jossey-Bass Publishers.
Joss, Simon, 1998. "Danish Consensus Conferences as a model of participatory technology assessment: an impact study of consensus conferences on Danish Parliament and Danish public debate." *Science and public Policy, 25*: 2–22.
———, 2002. "Toward the Public Sphere—Reflections on the Development of Participatory Technology Assessment," *Bulletin of Science, Technology and Society, 22*: 220–31.
Kanigel, Robert, 1997. *The One Best Way: Frederick Winslow Taylor and the Enigma of Efficiency.* Baltimore: Penguin Books.
Kaplan, Fred, 2013. *The Insurgents: David Petraeus and the Plot to Change the American Way of War*. New York: Simon and Schuster.
Kaple, Deborah, 1994. *Dream of a Red Factory: The Legacy of High Stalinism in China*. New York and Oxford: Oxford University Press.

Kato, Shinichi, 2006. "Upgrading Narrow Gauge Standards," *Japanese Railway History, 6*: 40–44.

Kidd, Benjamin, 1898. *The Control of the Tropics.* New York: The Macmillan Company.

Kijne, Hugo, and J. C. Spender, eds., 1996. *Scientific Management: Frederick Winslow Taylor's Gift to the World?* Boston, Dordrecht, and London: Kluwer Academic Publishers.

Kleinman, Daniel, ed., 2000. *Science, Technology, and Democracy.* Albany: State University of New York Press.

Klitgaard, Robert, and Paul Light, eds., 2005. *High Performance Government: Structure, Leadership, Incentives.* Santa Monica: RAND Corporation.

Kwok, D. W. Y., 1971. *Scientism in Chinese Thought, 1900–1950.* New York: Biblio and Tannen.

Kwong, Julia, 1985. "Changing Political Culture and Changing Curriculum: An Analysis of Language Textbooks in the People's Republic of China." *Comparative Education, 21*: 197–208.

Layton, Edwin, 1962. "Veblen and the Engineers," *American Quarterly, 14:* 64–72.

———, 1986. *The Revolt of the Engineers: Social Responsibility and the American Engineering Profession*, 2nd edition. Baltimore: Johns Hopkins University Press.

Le Corbusier, 1923. *Towards a New Architecture.* 1986 reprint, New York: Dover.

Levitt, Norman, and Paul Gross, 1994. "The Perils of Democratizing Science," *The Chronicle of Higher Education,* October 5. B1–B2.

Lewis, Sinclair, 1922. *Babbitt.* New York: Harcourt, Brace and Company.

Leydens, John, 2013. "Integrating Social Justice into Engineering Education from the Margins: Guidelines for Addressing Sources of Faculty Resistance to Social Justice Education," in Lucena, 2013: 179–200.

Li, Cheng and David Bachman, 1989. "Localism, Elitism, and Immobilism: Elite Formation and Social Change in Post-Mao China," *World Politics, 42*: 64–94.

Li, Cheng and Lynn White, 1988. "The Thirteenth Central Committee of the Chinese Communist party: From Mobilizers to Managers," *Asian Survey, 28:* 371–99.

———, 1990. "Elite Transformation and Modern Change in Mainland China and Taiwan: Empirical Data and the Theory of Technocracy," *The China Quarterly, #121:* 1–35.

———, 1991. "China's Technocratic Movement and the World Economic Herald," *Modern China, 17*: 342–88.

Litterer, Joseph, 1961. "Systematic Management: The Search for Order and Integration," *Business History Review, 35*: 461–76.

Loeb, Harold, (1933) 1996. *Life in a Technocracy: What It Might Be Like.* Syracuse, NY: Syracuse University Press.

Lucena, Juan; Carl Mitcham; John Leydens; J. Munakata-Marr; j. Straker; and M. Simoes, 2007. "Theory and Practice of Humanitarian Ethics in Graduate Engineering Education," In *2007 ASEE Annual Conference and Exposition, Conference Proceedings.*

Lucena, Juan; Jen Schneider; and Jon Leydens, 2010. *Engineering and Sustainable Community Development.* San Rafael, CA: Morgan and Claypool.

Lucena, Juan, ed., 2013. *Engineering Education for Social Justice: Critical Explorations and Opportunities.* Berlin: Springer.

Lynch, William T., 1997. "Teaching Engineering Ethics in the United States," *IEEE Technology and Science Magazine, Winter 1997/1998*: 27–36.

Mahoney, Michael J., 1989. "Scientific Psychology and Radical Behaviorism," *The American Psychologist, 44*: 1372–377.

Maier, Charles, 1970. "Between Taylorism and Technocracy: European Ideologies and the Vision of Industrial Productivity in the 1920s," *Contemporary History, 5:* 27–61.

———, 1975. *Recasting Bourgeois Europe: Stabilization in France, Germany, and Italy in the Decade after World War I.* Princeton: Princeton University Press.

Majone, Giandomenico, 1998. "Europe's 'Democratic Deficit': The Question of Standards," *European Law Journal, 4*: 5–28.

Marcuse, Herbert, 1964. *One Dimensional Man: Studies in Ideology of Advanced Industrial Society.* London: Routledge.

Martin, Brian, ed., 1996. *Confronting the Experts.* Albany: State University of New York Press.

May, Glenn, 1980. *Social Engineering in the Philippines* Westport: Greenwood Press.

Mazur, Allan, 1977. "Public Confidence in Science," *Social Studies of Science, 7*: 125.

McAvoy, Gregory, 1999. *Controlling Technocracy: Citizen Rationality and the NIMBY Syndrome.* Washington, DC: Georgetown University Press.

McDonald, Bruce D. III, 2010. "The Bureau of Municipal Research and the Development of a Professional Public Service," *Administration and Society, 42*: 815–35.

McKean Roland, 1953. *Sub-optimization criteria and operations research.* Santa Monica: RAND.

———, 1958. *Efficiency in Government Through Systems Analysis: With Emphasis on Water Resource Development.* New York: John Wiley.

Medvedev, Zhores A., 1969. *The Rise and Fall of T. D. Lysenko*. New York: Columbia University Press.

Merelman, Richard, 2000. "Technological Cultures and Liberal Democracy in the United States," *Science, Technology, and Human Values, 25*: 167–94.

Merkle, Judith, 1980. *Management and Ideology: The Legacy of the international Scientific Management Movement.* Berkeley and Los Angeles: University of California Press.

Meynaud, Jean, 1968. *Technocracy*. New York: The Free Press.

Mihalasky, John, 1996. "Scientific Management in Central Eastern Europe—Czechoslovakia, Hungary, and Poland," in Kijne and Spender, 1996.

Minkler, Meredith, 2010. "Linking Science and Policy Through Community-Based Participatory Research to Study and Address Health Disparities," *American Journal of Public Health, 100 (Supplement 1)*: S81–S87.

Monthoux, Pierre Guillet de, 1991. "Modernism and the Dominating Firm—On the Managerial Mentality of the Swedish Model," *Scandinavian Journal of Management, 7*: 27–40.

Moravcsik, Andrew, 2004. "Is there a 'Democratic Deficit in World Politics? A Framework for Analysis," *Government and Opposition, 39*: 336–63.

Morgan, Mary, 2003, "Economics," in Theodore Porter and Dorothy Ross, eds., *The Modern Social Sciences, Volume 7 of The Cambridge History of Science.* Cambridge: Cambridge University Press, pp. 275–305.

Morgan, Stephen L., 2006. "Transfer of Taylorist ideas to China, 1910–1930s," *Journal of Management History, 12*: 408–24.

Morris, William, 1906 (1890). *News from Nowhere, or An Epoch of Rest, Being Some Chapters from A Utopian Romance*. New York: Longmans and Green.

Morse, Philip M. and George E. Kimball, 1951. *Methods of Operations Research.* New York: John Wiley.

Muller, Edward N., 1985. "Dependent Economic Development, Aid Dependence on the United States, and Democratic Breakdown in the Third World," *International Studies Quarterly, 29*: 445–69.

Nash, George H., 1996. *The Life of Herbert Hoover: Master of Emergencies: 1917–1918*. New York: Norton.

National Academy of Engineering, 2008. *Engineering, Social Justice, and Sustainable Community Development*. Washington, DC: Center for Engineering Ethics and Society.

National Science Foundation, 2012. *Science and Engineering Indicators.* Washington, DC: U.S. Government Printing Office.

Nelkin, Dorothy, 1971. "Scientists in an Environmental Controversy," *Science Studies, 1*: 245–261.

———, 1975. "The Political Impact of Technical Expertise," *Social Studies of Science, 5*: 35–54.

———, 1981. *The Atom Besieged: Extra Parliamentary Dissent in France and Germany.* Cambridge, MA: MIT Press.

Nelson, Daniel, 1980. *Frederick W. Taylor and the Rise of Scientific Management*. Madison: University of Wisconsin Press.

———, ed., 1992. *A Mental Revolution: Scientific Management since Taylor.* Columbus: Ohio State University Press.

New York Times, 1962. "Computers Use in Arms Plans Splits Pentagon," March 28.

Newberry, Byron; Katherine Austin; William Lawson; Greta Gorsuch; and Thomas Darwin, 2011. "Acclimating International Graduate Students to Professional Engineering Ethics," *Science and Engineering Ethics, 17*: 171–94.

Nieburg, H.L., 1966. *In the Name of Science.* Chicago: Quadrangle Books.

Nieusma, Dean, 2013. "Engineering, Social Justice, and Peace: Strategies for Educational and Professional Reform," in Lucena, 2013: 19–40.

Nilsson, Tommy, 1996. "Lean Production and White Collar Work: The Case of Sweden," *Economic and Industrial Democracy, 17*: 447–72.

Noble, David, 1977. *America By Design: Science, Technology, and the Rise of Corporate Capitalism.* New York: Knopf.

Novick, David, 1962. *Program Budgeting: Long Range Planning in the Department of Defense.* Santa Monica: The RAND Corporation.

Nutting, P.G., 1917. "Organized Knowledge and National Welfare," *Science, 46*: 247–52.

O'Fallon, Liam; Geraldine Wolfle; David Brown; Allen Dearry; and Kenneth Olden, 2003. "Strategies for Setting a National Research Agenda that Is Responsive to Community Needs," *Environmental Health Perspectives, 111*: 1855–860.

Olson, Richard, 1982. *Science Deified and Science Defied: The Historical Significance of Science in Western Culture, from the Bronze Age to the Beginnings of the Modern Era ca. 3500 B.C. to ca. A.D. 1640.* Berkeley and Los Angeles: University of California Press.

———, 2008. *Science and Scientism in Nineteenth Century Europe*. Urbana-Champaign: University of Illinois Press.

———, 2010. *Technology and Science in Ancient Civilizations*. Santa Barbara: ABC-Clio.

Oreskes, Naomi, and Erik Conway, 2011. *Merchants of Doubt: How a Handful of Scientists Obscured the Truth on Issues from Tobacco Smoke to Global Warming.* New York: Bloomsbury Press.

Ottinger, Gwen, 2013. *Refining Expertise: How Responsible Engineers Subvert Environmental Justice Challenges.* New York: New York University Press.

Peci, Alketa, 2009. "Taylorism in the Socialism that Really Existed," *Organization, 16:* 289–301.

Peng, Khor Kok, 1988. "Science and Development: Underdeveloping the Third World," in Sardar 1988.

Picone, Antoine, 2007. "French Engineers and Social Thought, 18th–20th Centuries: An Archeology of Technocratic Ideals," *History and Technology, 23:* 197–208.

Pinchot, Gifford, 1910. *The Fight for Conservation*. New York: Doubleday.

Pinker, Steven, 2013, "Science is Not Your Enemy," *New Republic, 244* (August 6): 28ff.

Polanyi, Michael, 1962. "The Republic of Science," *Minerva, 1:* 54–73.

Pooley, Jefferson and Mark Solovey, 2010. "Marginal to the Revolution: The Curious Relationship between Economics and the Behavioral Sciences Movement in Mid-Twentieth-Century America," *History of Political Economy, 42*: 199– 233.

Prados, John W., ed., 2007. *A Proud Legacy of Quality Assurance in the Preparation of Technical Professionals: ABET 75th Anniversary Retrospective.* Baltimore: ABET Inc.

President's Research Committee on Social Trends, 1933. *Recent Social Trends in the United States.* New York: McGraw Hill Book Company.

Pursell, Carroll, 1995. *The Machine in America: A Social History of Technology.* Baltimore: Johns Hopkins University Press.

Putnam, Robert, 1977. "Elite Transformation in Advanced Industrial Societies," *Comparative Political Studies, 10:* 383– 412.

Radaelli, Claudio, 1999a. "The Public Policy of the European Union: Whither Politics of Expertise," Paper delivered at the ECSA International Conference, at Pittsburgh, June 2–June 5.

———, 1999b. *Technocracy in the European Union*, London: Routledge.

———, 2007. "Regulatory Impact Assessment, Political Control and the Regulatory State." Paper delivered at the 4th General Conference of the European Consortium for Political Research, Pisa, Italy, September 6–September 8.

RAND Corporation, 2009. *Standards for High-Quality Research and Analysis.* Santa Monica: RAND Corporation

———, 2013. RAND Corporation Official Web site at http://www.rand.org/about/people.html.

Rassweiler, Anne, 1988. *The Generation of Power: The History of Dneprostroi.* New York: Oxford University Press.

Ricks, Thomas E., 2012. *The Generals: American Military Command from World War II to Today*. New York: The Penguin Press.

Riley, Donna, 2013. "Power. Systems. Engineering. Traveling Lines of Resistance in Academic Institutions," in Lucena, 2013: 41–63.

Rosen, Christine, 2004. *Preaching Eugenics: Religious Leaders and the American Eugenics Movement.* New York and Oxford: Oxford University Press.

Rosenberg, Nathan, 1970. "Economic Development and the Transfer of Technology: Some Historical Perspectives," *Technology and Culture, 11*: 550–75.

Rostow, Walter W., 1990. *The Stages of Economic Growth: A Non-Communist Manifesto. Third edition.* Cambridge: Cambridge University Press.

Runneby, Nils, 1978. "Americanism, Taylorism, and Social Integration," *Scandinavian Journal of History, 3: 21–46.*

Ruskin, John. [1849] 1891. *The Seven Lamps of Architecture.* New York: John Wiley and Sons.

Sakellariou, Nicholas, 2013. "A Framework for Social Justice in Renewable Energy Engineering," in Lucena, 2013: 243–67.

Samuelson, Franz, 1981. "Struggle for Scientific Authority: The Reception of Watson's Behaviorism, 1913–1920," *Journal for the History of the Behavioral Sciences, 17:* 399–425.

Sardar, Ziauddin, ed., 1988. *The Revenge of Athena: Science, Exploitation, and the Third World*. London: Mansell.

Sarewitz, Daniel, 1996. *Frontiers of Illusion: Science, Technology, and the Politics of Progress.* Philadelphia: Temple University Press.

Sasaki, Satoshi, 1992. "The Introduction of scientific Management by the Mitsubishi Electric Engineering Co. and the Formation of an Organized Scientific Management Movement in Japan in the 1920s and 1930s," *Business History, 34, #2:* 12–27.

Schachter, Hindy Lauer, 1989. *Frederick Taylor and the Public Administration Community: A Reevaluation.* Albany: State University of New York Press.

Scientific American, 2010. "In Science We Trust: Poll Results on How You Feel about Science," September 22. Accessed at http://www.scientificamerican.com/article.cfm?id=in-science-we-trust, on September 9, 2013.

Schön, Lennart, 2012. *An Economic History of Modern Sweden*. London and New York: Routledge.

Scitovsky, Tibor, 1985. "Economic Development in Taiwan and South Korea: 1965–81," *Food Research Institute Studies, 19*: 215–64.

Sclove, Richard, 2000. "Town Meetings on Technology: Consensus Conferences as Democratic Participation." pp. 33–48, in Kleinman, 2000.

Scott, James C., 1998. *Seeing Like a State: How Certain Schemes to Improve The Human Condition Have Failed*. New Haven, CT: Yale University Press.

Segal, Howard, 1985. *Technological Utopianism in American Culture.* Chicago: University of Chicago Press.

Simon, Herbert, 1960. *The New Science of Management Decision.* New York: Harper.

Smith, Bruce L. R., 1966. *The Rand Corporation: Case Study of a Nonprofit Advisory Corporation.* Cambridge, MA: Harvard University Press.

Sochor, Zenovia, 1981. "Soviet Taylorism Revisited," *Soviet Studies, 33:* 246–64.

Solovey, Mark, and Hamilton Cravens, eds., 2012. *Cold War Social Science: Knowledge Production, Liberal Democracy, and Human Nature*. New York: Palgrave Macmillan.

Sorell, Tom, 1991. *Scientism: Philosophy and the Infatuation with Science*. London: Routledge.

Soubotina, Tatyana and Kathleen A. Sheram, 2000. *Beyond Economic Growth: Meeting the Challenges of Global Development.* Washington, DC: The World Bank.

Speer, Albert, 1970. *Inside the Third Reich: Memoirs by Albert Speer*. New York: The Macmillan Company.

Stokals, D; K.L. Hall; and A.L. Vogel, 2013. "Transdisciplinary Public Health: Core Characteristics, Definitions, and Strategies for Success," In Joshu and McBride, 2013: 3–30.

Strang, David and Young-Mi Kim, 2003. "The Diffusion and Domestication of Managerial Innovations: The Spread of Scientific Management, Quality Circles, and TQM between the US and Japan," pre-publication document available from Google Scholar.

Sullivan, William; Matthew S. Rosin; Lee S. Shulman; and Gary Fenstermacher, 2008. *A New Agenda for Higher Education: Shaping a Life of the Mind for Practice.* New York: Jossey-Bass.

Suttmeier, Richard, 2007. "Engineers Rule, OK?" *New Scientist.* November: 71–73.

Taylor, Blaine, 2010. *Hitler's Engineers: Fritz Todt and Albert Speer—Master Builders of the Third Reich.* Philadelphia: CASEMATE.

Taylor, Frederick Winslow, 1911 (reprint, 1967). *The Principles of Scientific Management.* New York: Norton.

———, 1919. (2nd edition. 1903 original), *Shop Management.* New York: Harper and Brothers.

Tesi, Francesca, 2007. "The Application of Taylorism in France: The Role of the Michelin Family in the Rationalization of French Work," *Business and Economic History on Line, 7:* 1–22.

Thackray, Arnold, ed., 1998. *Private Science: Biotechnology and the Rise of the Molecular Sciences.* Philadelphia: University of Pennsyvania Press.

Tharp, Lawrence, 1970. "Columbia Radicals Fight for their Academic Freedom," *U. R. P. E. Newsletter, vol. 2, #4:* 1–3.

Tolon, Kayla, 2012. "Futures Studies: A New Social Science Rooted in Cold War Strategic Thinking," pp. 45–62. In Solovey and Cravens, 2012.

Truman, Harry S., 1949. *Innaugural Address.* Found at http://www.trumanlibrary.org/whistlestop/50yrarchive/inagural20jan1949.ht. Accessed on November 13, 2013.

Tsutsui, William, 1998. *Manufacturing Ideology: Scientific Management in Twentieth-Century Japan.* Princeton, NJ: Princeton University Press.

———, 2001. "The Way of Efficiency: Ueno Yoichi and Scientific Management in Twentieth Century Japan," *Modern Asian Studies, 35 #2:* 441–67.

Tyack, David B., 1974. *The One Best System: A History of American Urban Education.* Cambridge, MA: Harvard University Press.

Tyagunenko, V.L., ed., 1976. *The Third World and Scientific and Technical Progress.* Moscow: Nauka Publishing House.

Vallianatos, Evaggalos, 2001. "All of Africa's Gods Are Weeping," *Race and Class, July-September,* 45–57.

———, 2006. *This Land os Their Land: How Corporate Farms Threaten the World.* Monroe, ME: Common Courage Press.

———, 2015. *Poison Spring: The Secret History of Pollution and the EPA.* New York: Bloomsbury Press.

Vitalis, Robert, 2010. "The Noble American Science of Imperial Relations and Its Laws of Race Development," *Comparative Studies of Society and History, 52*: 909–38.

Vonnegut, Kurt, 1952. *Player Piano*. New York: Dell (reprint 1981).

Wadwa, Vivek, Gary Cereffi, Ben Rissing, and Ryan Ong, 2007. "Where the Engineers Are: Seeing Through Preconceptions: A Deeper Look at China and India," *Issues in Science and Technology on Line,* Spring: 1–15.

Wahlke, John C., 1979. "Pre-behavioralism in Political Science," *The American Political Science Review, 73:* 9–31.

Waldo, Dwight, 1984. *The Administrative State: A Study of the Political Theory of American Public Administration, 2nd edition*. New York: Holmes and Meier.

Wang, Jessica, 1999. "Merton's Shadow: Perspectives on Science and democracy since 1940," *Historical Studies in the Physical and Biological Sciences," 30:* 279–306.

Wang, Zuoyue, 2008. *In Sputnik's Shadow: The President's Science Advisory Committee and Cold War America.* New Brunswick, NJ: Rutgers University Press.

Warner, Malcolm, 1994. "Japanese Culture, Western Management: Taylorism and Human Resources in Japan," *Organization Studies, 15:* 509–33.

Weber, Max, 1946. *From Max Weber*. Ed. Hans Gerth and C. Wright Mills. New York: Oxford University Press.
Wieseltier, Leon, 2013. "Crimes Against Humanities: How Science Wants to Invade the Liberal Arts. Don't Let it Happen," *New Republic, 244* (September 3).
Wilson, David Sloan, 2002. *Darwin's Cathedral: Evolution, Relgion, and the Nature of Society.* Chicago: University of Chicago Press.
Wilson, Gordon, 2006. "Beyond the Technocrat? The Professional Expert in Development Practice," *Development and Change, 37*: 501–23.
Wilson, Woodrow, 1887. "The Study of Public Administration," *Political Science Quarterly 2*: 197–222.
Wolfe, Audra J., 2013. *Competing with the Soviets: Science, Technology, and the State in Cold War America.* Baltimore: Johns Hopkins University Press.
Wrege, Charles D. And Marie Stotka, 1978. "Cooke Creates an Classic: The Story behind F. W. Taylor's *Principles of Scientific Management.*" *The Academy of Management Review, 3*: 736–49.
Xiao Gongquin, 2003. "The Rise of the Technocrats," *Journal of Democracy, 14:* 60–65.
Xiaowei Zang, 1993. "The Fourteenth Central Committee of the CCP: Technocracy or Political Technocracy?" *Asian Survey, 33*: 787–803.
———, 1999. "The Consolidation of Political Technocracy in China: The Fourteenth and Fifteenth Central Committees of the CCP," *Journal of Communist Studies and Transition Politics, 15:* 101–13.
Young Nam Cho, 2008. "Elite Politics and the 17th Party Congress in China: changing norms amid continuing questions," *The Korean Journal of Defense Analysis, 20:* 155–68.
Zak, Paul, 2008. "The Neurobiology of Trust," *Scientific American*: 88–95.

Index

About the Author

Richard Olson received his PhD in history of science from Harvard in 1967 and is currently professor of history emeritus at Harvey Mudd College as well as adjunct professor of history at Claremont Graduate University. In addition to writing three previous volumes on the history of scientism (*Science Deified and Science Defied, vol. 1*, University of California Press, 1982; *Science Deified and Science Defied, vol. 2*, University of California Press, 1990; and *Science and Scientism in Nineteenth Century Europe*, University of Illinois Press, 2008), he has written extensively on the interactions between science and religion and on the early development of the social sciences.

www.ingramcontent.com/pod-product-compliance
Ingram Content Group UK Ltd.
Pitfield, Milton Keynes, MK11 3LW, UK
UKHW041124071026
15587UKWH00009B/188

9 781498 525701